Android™ Design Patterns

Android™ Design Patterns

INTERACTION DESIGN SOLUTIONS FOR DEVELOPERS

Greg Nudelman

WILEY

Android™ Design Patterns: Interaction Design Solutions for Developers

Published by
John Wiley & Sons, Inc.
10475 Crosspoint Boulevard
Indianapolis, IN 46256
www.wiley.com

Copyright © 2013 by John Wiley & Sons, Inc., Indianapolis, Indiana
Published simultaneously in Canada

ISBN: 978-1-118-39415-1
ISBN: 978-1-118-41755-3 (ebk)
ISBN: 978-1-118-43934-0 (ebk)
ISBN: 978-1-118-65058-5 (ebk)

Manufactured in the United States of America

10 9 8 7 6 5 4 3 2 1

For general information on our other products and services please contact our Customer Care Department within the United States at (877) 762-2974, outside the United States at (317) 572-3993 or fax (317) 572-4002.

Wiley publishes in a variety of print and electronic formats and by print-on-demand. Some material included with standard print versions of this book may not be included in e-books or in print-on-demand. If this book refers to media such as a CD or DVD that is not included in the version you purchased, you may download this material at http://booksupport.wiley.com. For more information about Wiley products, visit www.wiley.com.

Library of Congress Control Number: 2012956395

Trademarks: Wiley and the Wiley logo are trademarks or registered trademarks of John Wiley & Sons, Inc. and/or its affiliates, in the United States and other countries, and may not be used without written permission. Android is a trademark of Google, Inc. All other trademarks are the property of their respective owners. John Wiley & Sons, Inc. is not associated with any product or vendor mentioned in this book.

For Katie and Juliette: Let your dream be the golden compass you live your life by.

ABOUT THE AUTHOR

Greg Nudelman believes in *designing what works*. His first experience with designing for mobile came when he joined the SkunkWorks team that created the original eBay mobile app that today generated more than $5 billion in revenue.

For more than 15 years, Greg helped craft cross-platform digital experiences for today's top Fortune 500 companies, non-profits, and startups: eBay, WebEx, Wells Fargo, Safeway/Vons, Cisco, IBM, Groupon, Associated Press, the U.S. Patent Office, and many others.

Greg is the author of *Designing Search: UX Strategies for eCommerce Success* (Wiley, 2011), which has a solid 5-star rating on Amazon. The book includes 19 perspectives from today's top names in search (a fact that Greg is particularly proud of).

Greg has contributed chapters and perspectives to the following publications:

- *Mobile Design Patterns* (Smashing Media, 2012)
- *The Mobile Book* (Smashing Media, 2013)
- *Designing the Search Experience*, Tony Russell-Rose and Tyler Tate (2013, Morgan-Kaufmann)
- *Search Analytics for Your Site,* Lou Rosenfeld (Rosenfeld Media, 2011)

Greg's work on storyboarding tablet transitions was featured recently in Rachel Hinman's *The Mobile Frontier* (Rosenfeld Media, 2012).

Greg has authored more than 30 industry articles on mobile and tablet design and digital design strategy for leading industry magazines: *Smashing Magazine*, *Boxes and Arrows*, *JavaWorld*, *ASP.NET Pro*, *UXmatters*, and *UXMagazine*.

He is a FatDUX, Rosenfeld Media, Wiley, and eConsultancy affiliate and workshop leader, and he has taught design workshops at Marquette University, HULT Business School, Associated Press, and Wells Fargo.

Greg is an internationally acclaimed speaker, with repeated appearances and sold-out workshops at leading industry events such as Adaptive Path's UXWeek, SXSW, MobX, IA Summit, WebVisions, Design4Mobile, Search Engine Summit, Enterprise Search Summit, Net Squared Conference, DrawCamp, and SketchCamp.

He is a co-founder of the UX SketchCamp movement with the landmark UX SketchCamp SF 2011 event. Greg's cross-platform design strategy consulting company, DesignCaffeine, Inc., is based in the San Francisco Bay Area.

CREDITS

Executive Editor
Robert Elliott

Project Editor
Charlotte Kughen,
The Wordsmithery LLC

Technical Editor
Ambrose Little

Production Editor
Christine Mugnolo

Copy Editor
San Dee Phillips

Editorial Manager
Mary Beth Wakefield

Freelancer Editorial Manager
Rosemarie Graham

Associate Director of Marketing
David Mayhew

Marketing Manager
Ashley Zurcher

Business Manager
Amy Knies

Production Manager
Tim Tate

**Vice President and Executive
Group Publisher**
Richard Swadley

**Vice President and Executive
Publisher**
Neil Edde

Associate Publisher
Jim Minatel

Project Coordinator, Cover
Katie Crocker

Compositor
Maureen Forys,
Happenstance Type-O-Rama

Proofreader
Nancy Carrasco

Indexer
Johnna VanHoose Dinse

Cover Image
Greg Nudelman

Cover Designer
Ryan Sneed

ACKNOWLEDGMENTS

To anyone who's never written a book it is difficult to imagine the blood, sweat, and tears required to finish one. I wish to acknowledge the generous help of my agent, Neil Salkind of Studio B as well as my fantastic team at Wiley: Charlotte Kughen and Ambrose Little. Any inaccuracies in the book are my own and no fault of theirs. Also, I'd like to thank Robert Elliott, from whose creative mind the idea for this book idea was initially born. I also want to acknowledge generous help provided by Kimberly Johnson, who helped decipher and sort out key Android visual design themes. Last, but definitely not least, I want to thank my family for their strong support and continual tolerance during this time of missed family commitments, forgotten appointments, and general mental fog surrounding the focused time of book writing.

CONTENTS AT A GLANCE

CONTENTS

FOREWORD

The first thing Greg told me when we met was, "You wrote the book I was working on," referring to the book *Mobile Design Pattern Gallery* that I had just released with O'Reilly Media (2012). I felt a little guilty at the time, but now I am glad I beat him to it. But not for the reasons you might think.

When I started the pattern gallery, I focused on identifying universal design patterns across the six major mobile platforms. Two years later, the industry is maturing, and only three big players are left, each with their own distinct patterns and principles. Universal patterns are still valuable, but more valuable yet are the deep dives into specific operating systems.

Greg saw this coming and decided to focus on the fastest growing platform, Android, and the most sophisticated release yet, Jelly Bean. His book meets mobile designers and developers where they are most comfortable and expertly guides them towards mastery of mobile user experience.

This book is more of a workshop than a reference book. Greg builds upon the universal design patterns for mobile devices and tablets and the Android UI design guidelines and takes the topic further into hands-on practical applications of the design principles. Each section covers fundamentals, warns of pitfalls and antipatterns, and then puts the lessons to the test by showing in detail how to redesign an existing app. You can and should bring this book to design sessions, and you should share it with your team. You will save countless hours solely from using the patterns in Chapter 7, "Search," and Chapter 8, "Sorting and Filtering." By reading and using Greg's entire book you will tremendously improve all aspects of the mobile experience you will be creating for your customers.

Bottom line, there is no other resource out there that goes to this level of depth on Android application design. I just hope Greg writes a Windows pattern book next.

Enjoy,

Theresa Neil
UX Designer, Start-up Advisor, Author/Speaker
Theresa Neil Interface Designs (www.theresaneil.com)

INTRODUCTION

Let me begin by answering some questions about the book you hold in your hands.

Why Mobile Computing?

> JIM RHODES: *You're not a soldier.*
> TONY STARK: *Damn right, I'm not—I'm an army.*
>
> —*Iron Man*, Marvel Studios, 2008

Mobile computing is the most game-changing development in human history. You live in the most exciting age—one of almost limitless potential, where your information, idea, product (in short, any *meme*) can reach virtually every person on the planet in a matter of days, if not minutes. And that is because no other modern technology has the reach and the potential of mobile computing. But simple penetration is not enough. The transformative force of mobile technology comes from the way it cradles people, empowers them to connect easier, make smarter decisions, and frees their minds in their soaring flight to go beyond the mundane.

With the coming of capable touch smartphones, the relationship with technology has shifted to that of intuitive digital assistant, an extra organ with super-human sensors—a true symbiosis that can best be described as a relationship of a Cyborg with his cybernetic components, or that of Tony Stark and his Iron Man suit.

Iron Man is my favorite metaphor because the suit is not a part of Tony, yet when he puts it on, he is one with the device. The Iron Man suit takes Tony's intention and transmutes it into action, on a grand scale, and without much effort on Tony's part (that is, without *cognitive friction*). At the end of the day, the Iron Man is just a man. Yet the power is always inside him, as it is in each of us. It takes this unique symbiosis with technology to truly enable and unleash that incredible power.

The mobile phone is our Iron Man suit. *The mobile experience, when executed well, is a cybernetic skeleton.* If you design and develop your app skillfully, your customers will *feel* protected and empowered in a way similar to how Tony Stark feels when he puts on his Iron Man suit.

Why Android?

Anyone following mobile space is aware that in the beginning Android had a few growing pains. (And that's putting it mildly.) Market fragmentation, overall confusion born out of a lack of focus and standards, and overly frequent updates all bear some of the blame. Yet like a professional prizefighter fueled by massive adrenaline and steroids, Android embraced these challenges head-on and managed to improve and evolve rapidly and grow market share faster than anyone thought possible.

As of this writing, the Android smartphone operating system was found on three out of every four smartphones shipped during the third quarter of 2012 (3Q12). According to the International Data Corporation (IDC) Worldwide Quarterly Mobile Phone Tracker, total Android smartphone shipments worldwide reached 136.0 million units, accounting for 75 percent of the 181.1 million smartphones shipped in the third quarter of 2012. The 91.5 percent year-over-year growth was nearly double the overall market growth rate of 46.4 percent (https://www.idc.com/getdoc.jsp?containerId=prUS23771812). With the release of Android 4.0 Ice Cream Sandwich, Android created a purely digital, business-like demeanor with a powerful core of a set of standards that work on virtually every device, while also dealing a left hook to fragmentation through a set of clever responsive design decisions for the structure of the menu and navigation scheme. All this new serious business sense comes wrapped up in a set of open standards and a well-evolved code base.

In short, in my humble opinion, the state of the Android ecosystem is now the perfect storm combining the factors for explosive near-term growth and long-term market dominance. If you have been working with Apple iOS, Windows Mobile, BlackBerry, and older Android OS, or if this is your first foray into the mobile space, today is the *perfect* time to look at designing and developing Android 4.0 apps.

Why This Book?

If you want your customers to feel as empowered when using your app as when they put on the Iron Man cybernetic exoskeleton, you need to unlock the patterns behind effective mobile design and apply them to your context. The book in your hands is the key to those patterns. Within these pages is everything you need to succeed in creating a great mobile experience.

Use What Works

This book is about what works: design patterns. A design pattern is a repeatable solution that helps resolve a particular problem within a specific context. But why do you need patterns—isn't reading the Android design docs enough? What makes design patterns uniquely effective is the way they communicate best practices *while addressing the complexities involved in real design problems.* As Christopher Alexander (the early pioneer of design patterns as formal ideas) says in his book *Timeless Way of Building* (Oxford University Press, 1979), the patterns make up the vocabulary of a design language that can be used to build things that are whole, complete, and alive (what he calls "the quality without a name").

In addition to helping you build usable apps, design patterns are intensely practical building blocks: they are small and can be learned and understood easily. You can combine patterns to create usable and delightful designs. Finally, patterns form the design language you can use to communicate simply and effectively.

Apply 58 Essential Android App Patterns

In Part 2 of the book, you discover all the patterns you need to create great interaction design and intuitive Information Architecture for your Android 4.0+ apps. There are 58 essential interaction design patterns for dealing with the most challenging aspects of the Android app design: welcome experience, home screen, navigation, search, sorting and filtering, data entry, and forms. The patterns in this book are designed to look beyond the obvious and build on the official Google documentation, allowing you to move smoothly from theory to practical applications. In addition, there are specific chapters covering key design patterns for mobile banking and dealing with the tricky aspects of tablet design.

Avoid Common Pitfalls with 12 Antipatterns

In addition to 58 patterns, there are 12 antipatterns, describing the most common mistakes to avoid in your quest for customer empowerment, delight, and enjoyment. Standalone antipatterns are mobile evolutionary dead-ends you want to steer clear of. Sometimes you also see the same antipattern icon used as part of the regular pattern. These are common pitfalls (lined with spikes on the bottom) that will catch the unwary. Read those carefully—often only a part of the screen or a specific interaction is called out as the antipattern and not the entire screen. The antipatterns and negative examples are marked with the symbol you see in the margin next to this paragraph.

Be Inspired by Innovative Ideas

In addition to helping you build a rock-solid design pattern foundation, this book gives you the confidence and inspiration to move beyond the tried-and-true patterns to create exciting innovative implementations from existing mobile ideas and interface components. You can explore experimental patterns (marked with the symbol you see in the margin next to this paragraph), which stretch the existing ideas and mobile status quo.

In the workshops I teach around the world, people often ask me: "Do experimental patterns work?" To answer this question, let me tell you a short story. In September 2010, I presented a pattern I called Immersive Navigation at Design4Mobile in Chicago. I suggested that the fold-out menu navigation used by games like Angry Birds can and should be adopted to more "serious" mobile applications like e-commerce, news, and social media. Many of those present were skeptical: Would this work? Can you even get these apps past Apple's stringent guidelines that require the use of the tab bar? To this I replied that the Apple tab bar is merely a set of training wheels, and that I believe the mobile consumer is ready to upgrade to the latest Harley-Davidson, which, at the time, caused quite an uproar.

Less than a year later, Facebook came out with a new fold-out navigation menu on the top-left corner. Other successful apps started using it too; Flipboard, for example, used the same pattern on the top-right corner. Today this pattern, known as the Drawer, is part of the standard Android 4.0 toolkit, and it's used by apps like Google Plus. I cannot, of course, claim credit for this development. I merely hope that I helped to provide another tiny push in the direction many talented people were already taking.

Mobile design moves at incredible, unprecedented speed. Experimental patterns described in this book are possible near-future design patterns that run slightly outside current mainstream mobile approaches. For those willing to try out new ideas, these experimental patterns represent incredible *opportunities* to stand out from the 700,000 apps currently available in Google Play, leap-frog the competition, and deliver uniquely engaging mobile experiences ("Google Says 700,000 Applications Available for Android," *Bloomberg Businessweek*, 29 October 2012, Retrieved 5 November 2011). *But please don't take my word for it!* Instead, I invite you to try out and customer-test the experimental patterns you like, to see if they will work for your particular project. I also invite you to use the ideas in this book as an inspiration to thinking outside the training wheels and building your own design approaches. As Eckhart Tolle so eloquently says in his timeless book, *The Power of Now* (New World Library, 2004), "Try it out and you will *be* the evidence."

Use a Complete Design Methodology

Patterns are the focus of this book. However, Part 1 describes the complete sticky-notes methodology for building effective, cheap prototypes and customer-testing them. Part I also includes an entire chapter on Android visual design principles and a case study demonstrating the hands-on use of the principles.

The book you hold in your hands is the practical, hands-on culmination of 14 years of designing and building digital products. In it I share the most effective methodologies I developed for mobile customer-centered design. But you don't just get one chapter on methodology! Instead, a customer-centric methodology for mobile design is woven into every pattern. Practically every one of the 58 patterns in this book is accompanied by a meticulous, detailed drawing of how this pattern or interface control would look when implemented using the sticky-notes methodology, *one you can use as a guide to create your own lean agile prototypes.* If you need help, don't hesitate to ping me and my team at www.AndroidDesignBook.com, the companion site for this book, where you can watch detailed videos of mobile usability testing and get all your questions answered. This book is all about *what works*, and I want to make sure you get the most value from these patterns by putting them to work yourself on your own project.

Design What Works

I am not an adoring foam-at-the-mouth Google groupie. I have far too many projects under my belt—projects that required compromise, out-of-the-box thinking, and striking innovation—to be devoted to a single idea or doctrine. I have also seen many so called "pure" projects of every kind fail miserably. Thus in this book you will see Android adaptations of great ideas from other mobile operating systems such as Apple iOS, Windows Mobile, and even (gasp!) BlackBerry.

I outline the unique capabilities of the new Android OS in Part 1 of the book, while focusing the bulk of the material on practical issues that come up in design projects and effective solutions for solving real-world challenges. In short, this book is the compendium of everything you need to design great-looking and great-performing modern Android apps. Now, if you are ready to start, let's do this thing!

What About the Code?

Glad you asked! After all, fantastic, intuitive design is all well and good, but you need to implement it at some point. *There is no code in this book.* Separating design from implementation was a deliberate decision because mobile design is a sophisticated endeavor with crushing constraints and pitfalls at every step of the way, so it took the entire book just to get through the design portion of the project.

To help you with coding your app, I have built a companion book site, www.androiddesignbook.com, especially to provide the complete support for Envision-Design-Build app life cycle. On the site there are more than 100 articles, numerous code examples, and mini-apps you can learn from and copy for your own purposes; regular design webinars, which deal with audience-posed design challenges; and a dedicated team of experts to answer your questions. Most important, there is a large supportive Android community to help you every step of the way. And an Android design certification program is being set up. You can *register free* with your e-mail simply by typing the code **DROIDRULES** into the space provided on the sign-up form.

I hope you join us!

How Should You Use This Book?

This book is meant to be a hands-on reference throughout the design and development life cycle of your Android app. Part 2 of the book is something you will refer to over and over again. However, I—along with the fantastic editors at Wiley—have spent considerable effort to write the book as a story you can read from the beginning to end. Antipatterns are usually at the beginning of the chapters that include them. Simpler patterns are found earlier in the chapter, and more complex experimental ideas are placed toward the end. General patterns are at the beginning of Part 2, and the more specific applications, such as mobile-banking and tablet-specific patterns, are at the end of the book.

If you have a specific question, by all means, start with the pertinent chapter and dig in! However, at some point (sooner better than later) you should read Part 1. It is meant as a brief, effective introduction to the Android 4.0 design and sticky-notes design methodology. Even if you consider yourself an expert, at the very least, be sure to read Chapter 1, which includes the AutoTrader redesign case study. It's a great place to get your feet wet before you dig into the patterns in Part 2.

Who Should Read This Book?

I come to the discipline of customer-centered design from the background of back-end Java software architecture and Oracle databases. Thus, the material is hands-on, and the intended audience for this book is anyone involved with designing and developing Android apps. The book is aimed at mid- to advanced-level practitioners. However, a determined beginner can get full value from the book by using the design methodology to design, experiment, and build up skills to turn into an Android design expert. However, from the standpoint of design, goals, and monetization models, this book will likewise greatly benefit product managers, project managers, visual designers, user researchers, and business people by providing a common vocabulary with which to discuss mobile design and development challenges and various practical approaches to solving them.

PART I

UX Principles and Android OS Considerations

CHAPTER 1

Design for Android: A Case Study

This book is about what works: design patterns. Design patterns in this book build on the official Google Android design guidelines by communicating best practices while addressing the complexities involved in real design problems. The official Android guidelines (available at http://developer.android.com/design/get-started/ui-overview.html) form the foundation; this book shows you how to bring these guidelines to life as complete solutions to real-world design challenges.

With this chapter, I am laying the foundation for the 58 patterns (and 12 antipatterns) in the book by providing a case study of an app that could benefit from a more refined design—the AutoTrader app. The appropriate patterns are referenced in each section of this chapter; feel free to flip to the relevant pages to explore design solutions in more detail.

The AutoTrader app is a typical example of a *straight port*, which is to say that it is basically an iOS app that was quickly and minimally made to work for Android. The following sections show you how to redesign this app for Android 4.0+ (Ice Cream Sandwich). The entire app isn't covered because this would be exceedingly tedious to write (and even more tedious to read). Instead, three representative screens are discussed: home screen with a search form, the search results screen, and the item detail screen. These three screens should give you a good idea of some unique and interesting aspects of the Android visual design and navigation, and they give you a taste of the interaction design patterns in this book. Think of this chapter as an appetizer for a rich smorgasbord of practical solutions waiting for you in Part 2 of the book.

Launch Icon

The first thing to look at is the launch icon. Most apps that do a straight port from iOS neglect the essential part of redesigning the launch icon. The Android launch icon design is not bound by the iOS square shape with rounded corners. Designers are encouraged to give their Android launch icons a distinctive outline shape. Take a look at the launch icons for Yelp and Twitter in Figure 1.1—these folks get it.

FIGURE 1.1: The Yelp and Twitter launch icons have distinctive shapes.

In contrast, AutoTrader, the app for the case study, did not take the time to customize its icon. Fortunately, this is often a simple modification. In the case of AutoTrader, one suggested redesign is included in Figure 1.2. You could use the letter "A" borrowed from the rebranded iOS app and remove the background fill to create a distinctive shape. You are not bound to use a part of the logo—for instance the icon could have been in the shape of a car or steering wheel. The eye more readily perceives the shape of the icon when it is different from other apps, so this enables AutoTrader customers to find the app more easily in a long list.

FIGURE 1.2: The initial AutoTrader launch icon isn't distinctive, so here's a redesigned icon.

Action Bars and Information Architecture

In general, action bars and the accompanying functions form the nerve center of an app and are important in the overall design. Unfortunately, the current design of the AutoTrader app leaves much to be desired on this front (which is what makes this such a killer case study).

Before

Look at the default home screen: the Car Search. The most emphasized menu function is Settings, which is prominently featured in the top-right corner (see Figure 1.3). That location is arguably the second-most important and prominent spot in the mobile UI (the most prominent spot on the screen is top left, occupied by a large logo).

Although it's admirable to try to feature the Settings function, I unfortunately could not imagine a single primary or secondary use case that involves this function. Especially because what is labeled as Settings is nothing more than the placeholder for lawyer-fluff such as the privacy policy, visitor agreement, and a button to e-mail feedback—hardly the essential functionality that the app needs to feature so prominently!

In contrast to the over-emphasized Settings button, the essential functions that need to be used, such as Find Cars, Find Dealers, Scan & Find, and My AutoTrader, are hidden in the older, Android 2.3-style navigation bar menu (see Figure 1.4).

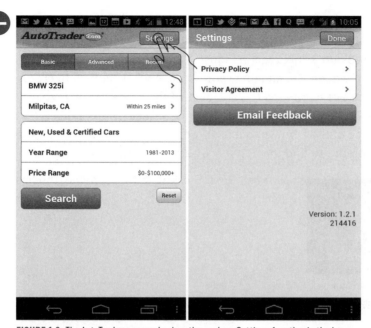

FIGURE 1.3: The AutoTrader app emphasizes the useless Settings function in the home screen design.

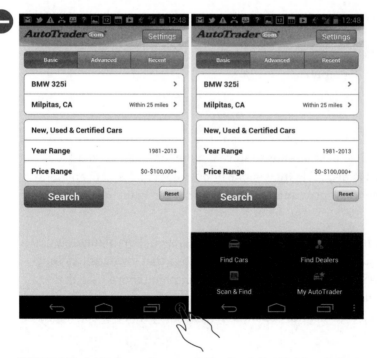

FIGURE 1.4: The AutoTrader app places essential functions in the old-style navigation bar menu, which is an antipattern.

The next section describes how the app could be redesigned according to the Android 4.0 guidelines to use action bars effectively and make the most important functions more prominent.

After

The first thing to fix is the style of the buttons. The rounder corners and bevels simply must go. So do the word-driven functions, such as Settings, in the action bar. In Android 4.0 the actions in the action bar are shown with icons, and the actions in the overflow menu are shown with text. Sticking to this scheme, the first suggestion is for the straightforward port of the old menu to the action bar, which might look like what's shown in Figure 1.5.

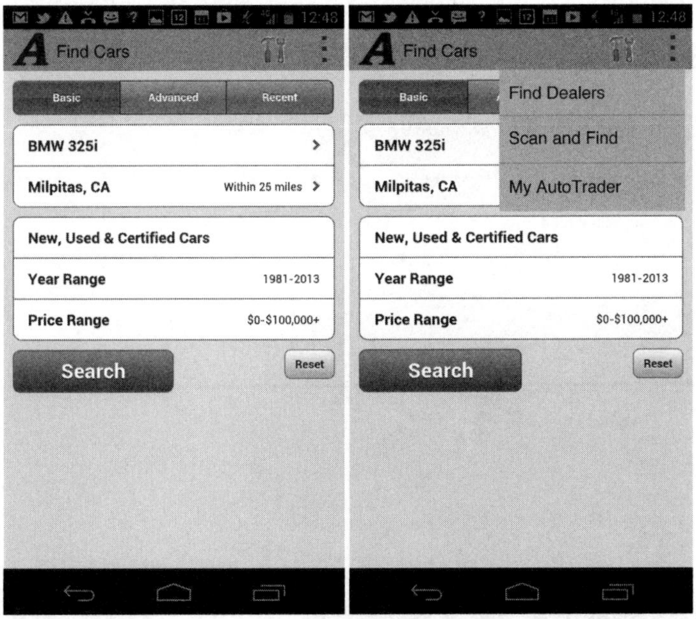

FIGURE 1.5: Version 1 is a straightforward port to Android 4.0 with settings and actions in the overflow menu.

In this version, the settings button has become the hammer and wrench icon, and the bottom navigation menu has been moved into the overflow function on the action bar. The giant company logo is replaced by the Android 4.0 style action bar icon (which matches the launch icon "A") and the screen title. (Note that according to the Android design guidelines, the screen title may not exceed 50 percent of the width of the screen, which is not a problem here; it's merely something you need to keep in mind.)

Unfortunately, as discussed in the "Before" section, these changes are not nearly enough. This basic redesign takes care of the Information Architecture (IA) port to Ice Cream Sandwich, but it does not take care of the inherent shortcomings of the app's current IA: Key functions such as Find Dealers and My AutoTrader are still hidden, and the Settings clearly does not go anywhere useful. Worse, placing Settings on the top bar would actually discourage exploration of the menu because if the customer discovers that the Settings function is basically pretty lame, that would send a strong signal that the other functions hidden in the overflow menu are even more useless. The design can be improved even more.

One possible approach would be to bubble up the Find Dealers and My AutoTrader options to the top action bar and remove the Settings to the overflow menu. Figure 1.6 shows how this might look.

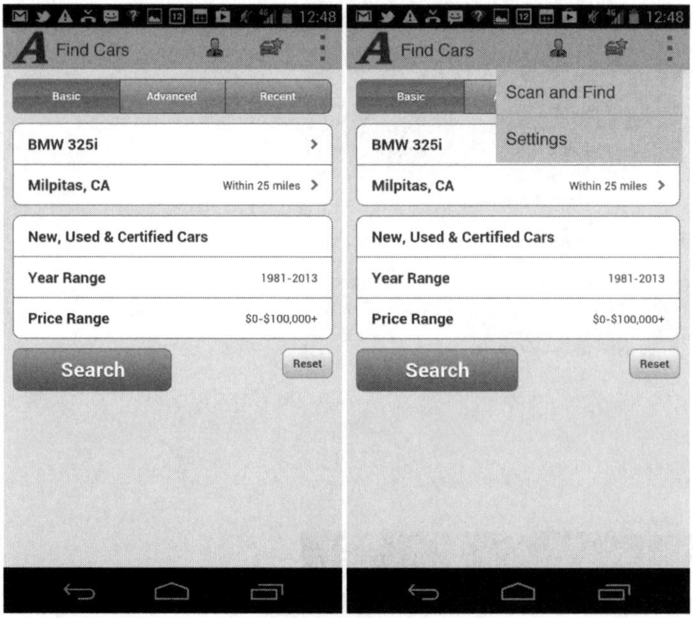

FIGURE 1.6: In Version 2, the more useful functions are on the top action bar and settings have been moved to the overflow menu.

This is an acceptable IA, and it is in line with the current Google Android recommendations for the Ice Cream Sandwich (4.0) and Jelly Bean (4.1) OS versions. However, it points out some key challenges with the implementation of the current UI specification of the action bars. For instance, on most devices, you cannot have more than a few functions on the action bar without taking up more than the recommended 50 percent of the available space. Furthermore, placing more actions on additional action bars robs the app of the vertical real estate it so desperately needs while also adding visual noise and complexity. This is not a small consideration that can be easily dismissed.

The main challenge is cognitive: Not every action can be easily represented with only an icon. For instance, in Figure 1.6, I use both of the original Find Dealers and My AutoTrader icons. While neither icon is bad, either one can be easily misinterpreted, as can most of the icons meant to represent complex or unusual actions. You could remove all the icons and place all of the actions in the overflow menu, but that solution is also far from ideal because it forces all the menu items to be solely text. When you use only text you miss out on the playful aspect that the icons bring to mobile computing, which is—at least for me—at the heart of what mobile navigation is all about. I have seen repeatedly that using icons and text *together* makes navigation most effective. When customers first learn the app, they rely on both aspects of the navigation. After using the app for a while, the icon often offers enough information scent to ensure recognition of the action behind it. So does Android offer a way to use both icons and text together?

Fortunately, the recent redesign of the Google Plus app points the way to use both icons and text by using a Drawer element (see Figure 1.7). The Drawer and other Swiss Army Navigation pattern techniques are covered in Chapter 13, "Navigation," so it's not necessary to cover them here. Suffice it to say that the Drawer user interface (UI) element enables both icons and text—the best of both worlds.

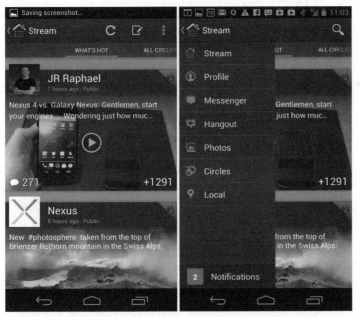

FIGURE 1.7: The Google Plus app design uses a Drawer menu that includes both text and icons.

The Android UI specification encourages the use of the Drawer element for top-level navigation if there are a number of views in the app that do not have a direct relationship with one another. This is exactly the case you have with Auto-Trader. The Car Search area of the app is different from the Find Dealer and My AutoTrader views, so placing these top-level navigation functions in the Drawer menu (shown in Version 3 of the redesign in Figure 1.8) makes a lot of sense. Scan & Find is a car search function, so it makes sense to make it contextual to the Car Search view. It is accessible with a single tap on the action bar. The useless (but, as the lawyers would argue, necessary) Settings function is the only one that hides in the overflow menu; it does not need to be accessible with one tap, so hiding it is the best strategy.

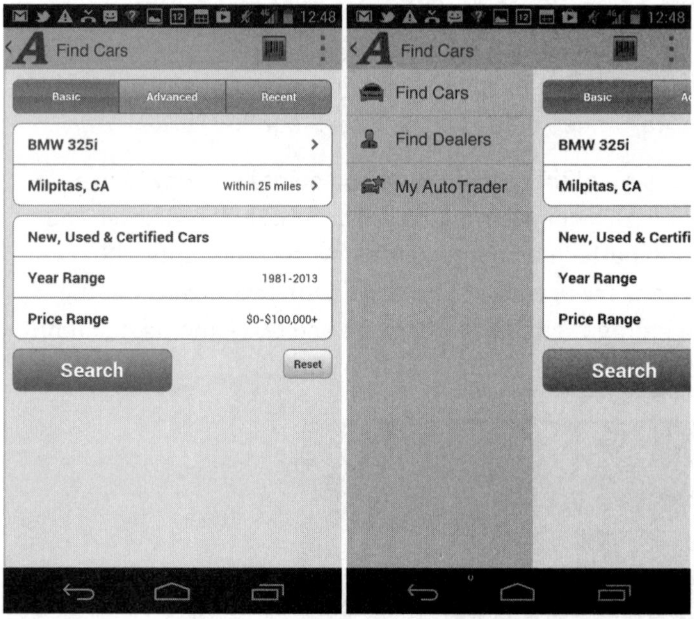

FIGURE 1.8: Version 3 is the recommended design for the AutoTrader app: a top-level redesign that uses a Drawer menu.

Version 3 is the preferred design. It strikes a good balance of showing both the icons and the text, while making the navigation accessible using a right-to-left swipe or a tap on the Up icon (left caret or ←). It also frees the top action bar for showing a good-sized, clear screen label. One recommended modification to the standard Android guidelines is a thin line all along the left edge of the screen that signals to customers that they can open the Drawer menu by swiping from right to left (as well as by tapping the Up icon).

The Android UI guidelines caution that the Drawer can be used only for top-level navigation, which means that while your customers are deep in the middle of the flow inside the Car Search view, they may be one or more steps away from accessing the additional views. The good news is that—with the global navigation out of the way—the action bar can include functions that are contextual to the page the customers are on, which is recommended by the Android design standards document.

Tabs

Tabs are an essential element of secondary navigation that can be used in the Android platform for a variety of applications. The Tabs pattern is covered in Chapter 8, "Sorting and Filtering."

The AutoTrader app uses the iOS style visual design rendering for the tabs with rounded corners and the "depressed" beveled look for the selected tab (see Figure 1.9).

Use a simple redesign of the control with end-to-end underline; no shadows, bevels, and rounded corners. The heavier "underline" element signals the selected tab. In this screen, there is just enough space for compact text labels (Basic | Advanced | Recent) so that is what is used in the suggested redesign.

What if the screen was smaller than could comfortably accommodate the full text in the tabs? Then the text labels would turn into corresponding icon-based tabs. As you read in Chapter 2, "What Makes Android Different," the scalability of running the interface on devices with smaller touch screens is a key differentiator for the Android OS. This scalability in turn dictates many of the basic visual design choices and guidelines.

FIGURE 1.9: The top shows the tabs in the AutoTrader app before a suggested redesign; the bottom is the suggested Android 4.0 treatment.

Dedicated Selection Page

Dedicated Selection Page is the primary pattern for selecting from a long list. It's covered in more detail in Chapter 12, "Mobile Banking." The AutoTrader app uses the iOS-style selection with the greater than sign, (right caret or →). (See the top of Figure 1.10.)

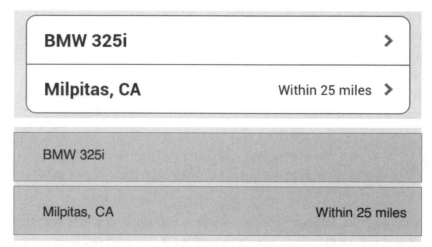

FIGURE 1.10: The top shows the link to the Dedicated Selection Page before redesign; the bottom is the suggested Android 4.0 treatment.

iOS uses the → to show row-based interactivity. In contrast, in the Android OS there is no indication of the underlying functionality. As discussed in Chapter 2, the concept of Tap Anywhere is an important one to the Android OS. If there is any reason to tap anything like a selector, the assumption is that it will carry the corresponding interactivity. Thus, visual design is implemented accordingly in a typical Spartan Android fashion, using a slightly darker row background and without the right caret.

Select Control

The Android platform comes with a full complement of touch-friendly controls that you can use on multiple screen sizes and device configurations. Ice Cream Sandwich comes fully equipped with touch sliders, a completely redesigned text entry, and a new dual-function wheel control, discussed in great detail in Chapter 10, "Data Entry." For this section, suffice it to say that the AutoTrader Android port still uses the iOS-style form controls and form section headers. The following sections describe how to redesign them, Android-style.

Before

The first thing to notice is the composite iOS-style wheel control that the customer uses in the AutoTrader app to select the year and price (see Figure 1.11).

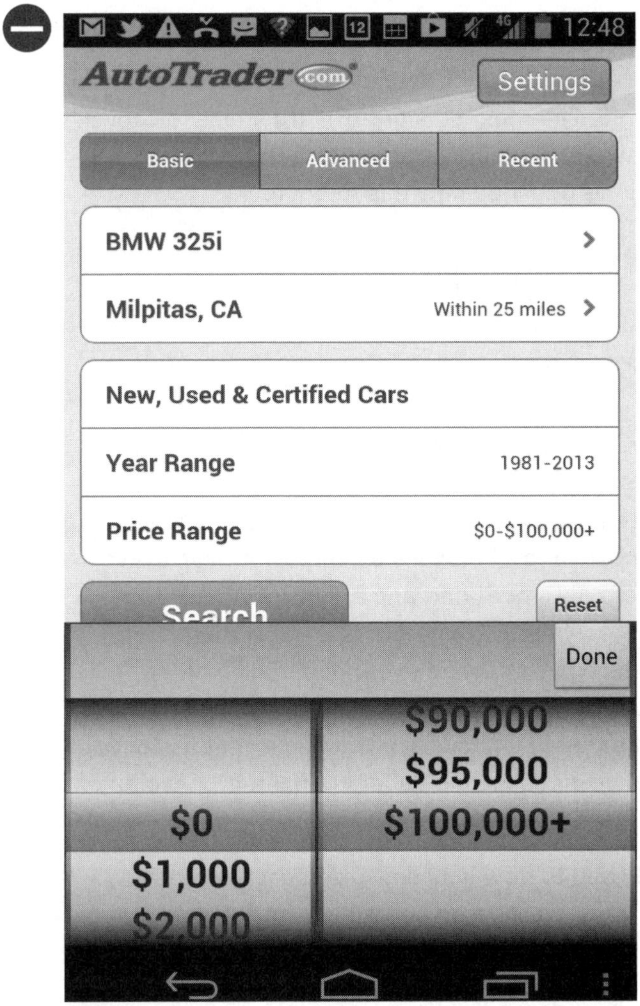

FIGURE 1.11: The AutoTrader app uses an iOS-style form for selecting the year and price.

The iOS control is a non-native wheel that needs to change to the Android-style controls. Check out a couple of ideas in the following "After" section.

Another important aspect of the form design is the rounded container for the entry fields with an iOS-style section header. As discussed in Chapter 2, Android uses the Mobile Space, Unbound visual style principle that removes any containers and boxes, especially those with rounded corners.

After

As discussed in Chapter 10, there are various touch-friendly ways to implement entry of the range of values in Android. The most straightforward one is to convert the combo wheel into min and max wheel controls (marked by a line with a small triangle on the right). Figure 1.12 shows one idea how this might look.

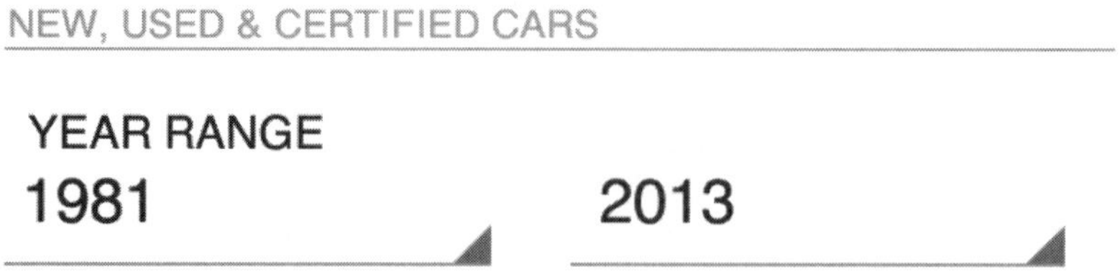

NEW, USED & CERTIFIED CARS

YEAR RANGE
1981   2013

FIGURE 1.12: The redesign uses native Android wheel controls and a section header.

Although wheel controls offer a decent solution, you can use a variety of other interaction design patterns. Chapters 10 and 11 discuss a composite Drop Down control, separate min and max Sliders, a Dual Slider, and a Slider with a Histogram experimental design pattern. Applying those patterns is not complicated, but it is a sophisticated endeavor, which is discussed in detail in Part 2 of the book. The patterns are designed especially to limit the cases of zero results in the search, such as picking a price range that is too low or having no inventory for a specific desired year, which is the topic discussed in detail in Chapter 9, "Avoiding Missing or Undesirable Results."

Forms in the Android 4.0/4.1 are allowed to flow in order to conform to a variety of screen heights and widths. Thus, instead of using containers for form sections, various parts of the form are separated from one other by simple headers. Native headers are in the all capital Roboto font (Helvetica is used in the figures) underlined by a thin separator line of the contrasting color.

Buttons

The AutoTrader app uses iOS-style buttons with rounded corners and bevels. The buttons are in two different heights and visual treatments, with lots of space in between, which makes the screen look rather lopsided. In addition, the buttons are positioned as Search/Reset, in other words in the OK/Cancel order (see Figure 1.13). As described in Chapter 11, "Forms," the preferred button orientation is the opposite: Cancel/OK, so the redesign flips the buttons from their "before" layout positions.

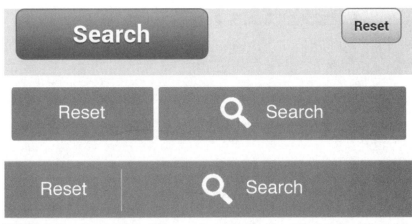

FIGURE 1.13: The current AutoTrader buttons are shown at the top; the redesign is in the middle; and an alternative of Android "tap areas" is at the bottom.

In contrast, the Android buttons are business-like: flat, with no gradient and just barely rounded corners. The preferred Android Cancel/OK button treatment is to turn them into square solid tap areas that occupy 100 percent of the width of the screen, with just a hint of a separator. Tap areas are discussed further in Chapter 2. I chose to emphasize the primary action button, Search, and make it easier to tap by making it larger and adding a magnifying glass icon.

Search Results

With the home screen and the IA redesigned, now turn your attention to the search results screen. Search results appear immediately after the customer searches with the home screen form, so this makes sense.

Before

Again, the search results screen for the AutoTrader app has been generally designed without adopting it to the Android Ice Cream Sandwich and Jelly Bean OS guidelines (see Figure 1.14). The screen uses mainly iOS standards, with a light mixture of Android thrown in. The screen uses three buttons: two with text and one with an icon. All three buttons use rounded corners and bevels. In addition, each result on the screen uses the iOS right caret → treatment. As seen previously on the home screen, the top app menu can be obtained by tapping the menu key in the navigation bar on the bottom of the device.

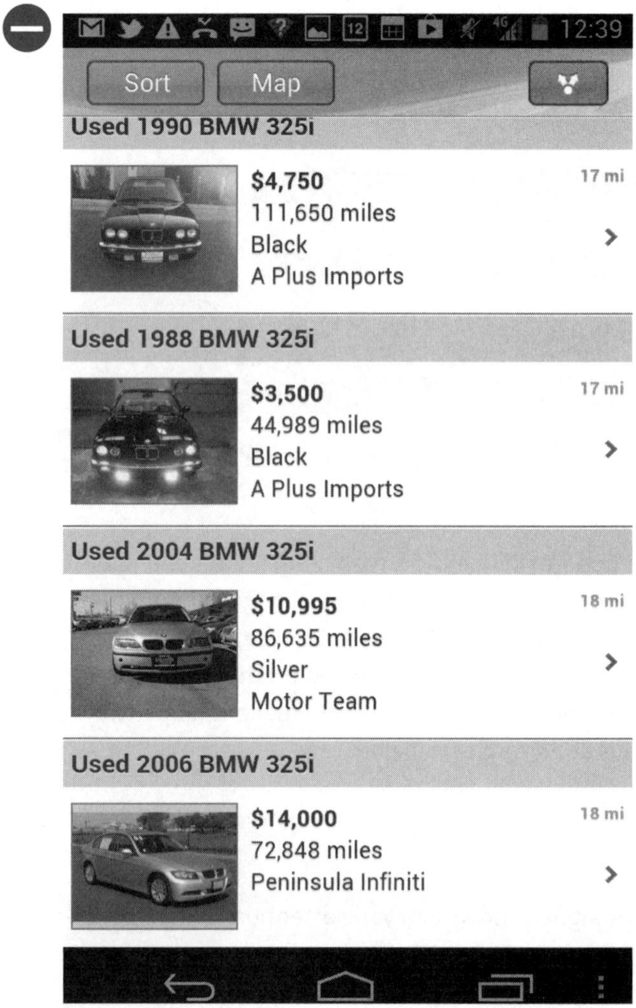

FIGURE 1.14: This is what the search screen looks like before redesign.

After

The Drawer top navigation menu is safely out of the way on the Search Results screen. Tapping the icon in the top-left corner navigates back to the home screen, where the customer can get to the top menu by tapping the same button again. This leaves space on the screen for contextual action buttons: Filter, Map, and Share (see Figure 1.15).

Starting with the Ice Cream Sandwich OS, the Share function has been a special use case because of the multiple built-in sharing functions. Thus you can implement Share as a standard drop-down menu per the Android UI design standards. The remaining Map and Filter buttons are implemented as the Android-style flat single-color icons, which are placed right on the action bar. This is one way that the map-list relationship can be implemented. Many more effective search and filtering patterns are discussed in Chapter 7, "Search," and Chapter 8.

FIGURE 1.15: This is Version 1 of the redesigned AutoTrader app search screen with Title Bar.

In addition to using the title bar treatment of the action bar, one other version is possible: a drop down control that can be used to select from multiple views: in this case from several different ways to sort the inventory. In this Version 2 of the redesign, the screen title (maker and model of the car) is displayed above the multiple views drop-down. Version 2 of the redesign is recommended because it adds key functionality by taking full advantage of the Android 4.0 capabilities (see Figure 1.16).

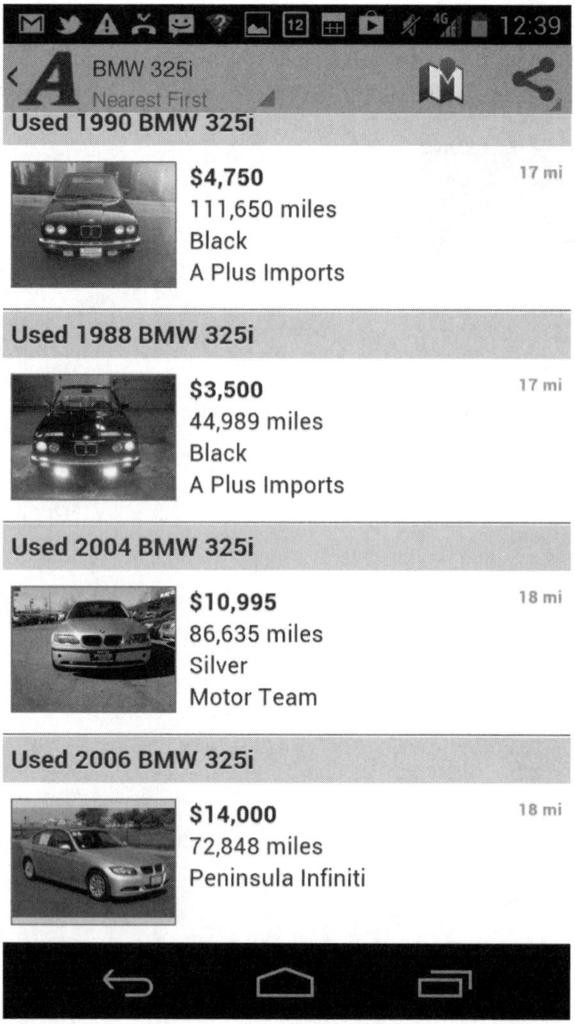

FIGURE 1.16: Recommended Version 2 of the AutoTrader app search screen redesign adds a Sort view selector.

The last thing to notice is the absence of the right caret → symbol in the search results. As mention earlier, touch space on the screen should be tap-enabled without having any special visual indicators. If an action, such as drilling down to the Result Detail screen, is valuable to the customer it should be enabled as a touch target without any external visual indicators. Speaking of which, it's time to pretend someone actually tapped the search result. The next section describes the third and final screen: Result Detail.

Result Detail

What happens when the customer drills down into a car detail screen? It turns out that the last screen offers many opportunities for the new Android redesign, from IA to tabs and buttons.

Before

The detail screen again includes many iOS elements (see Figure 1.17). As discussed earlier, tabs are text-based and have bevels and rounded corners. Similar to the search results screen, there is another Share button; however, in this screen it occurs twice: once on the top menu bar and once inside the screen in the guise of a Save/Share button. This is confusing.

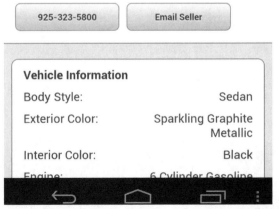

FIGURE 1.17: This is what the original AutoTrader app Result Detail screen looks like before redesign.

Other actions include call and e-mail the seller, yet no primary button can be identified in the entire set—it is not clear what the app wants the customer to do first. The rest of the screen is laid out using containers with rounded corners; those have to go, of course.

Most important, the only way to navigate to the next item in the list is to "pogostick": Press the Back button to return to the search results screen and then select a different detail screen. (*Pogosticking* is a navigation antipattern described in Chapter 13.)

After

In the redesigned version of the screen, continue with the simple Up navigation by removing the global navigation functionality. But where should the action buttons go, and what's the best way to identify the primary action that's most likely to be taken by the customer? It's easy to understand one way to implement this solution in the Android OS 4.0 by looking at the native Android Gmail app detail screen (see Figure 1.18).

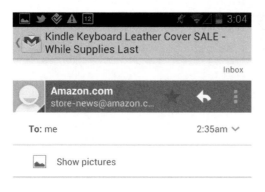

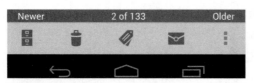

FIGURE 1.18: The Gmail app uses the Swipe Views control in the Result Detail screen.

To reduce pogosticking, the native mail app uses a clever Android OS interface control, Swipe Views, to make navigation more efficient. This control enables the customer to swipe from right to left to get to the next result detail. This functionality is shown to the customer using a thin, dark line on the bottom on the screen that states "2 of 133". Although this function works, in testing, the discoverability was poor. So for the redesign of the Android AutoTrader app, you should use a brief onscreen overlay tutorial as described in Chapter 5, "Welcome Experience," or an animated transition Watermark pattern described in Chapter 13 to highlight the Swipe Views control for the customer and improve discoverability of this important feature. Regardless of the introduction you choose to display, after the customer learns the action, the tutorial is no longer necessary and can be suppressed, so these patterns aren't shown here.

The swiping action in some applications is used to navigate between tabs. Because you want to preserve the swipe-to-next action to navigate to the next item detail, use the tap-only tabs on the top of the page, as shown in Figure 1.19.

FIGURE 1.19: This is how you can redesign the AutoTrader detail page.

The primary and secondary contextual actions can now be placed in the action bar. Because there are only three actions on the car detail page, you need only a single action bar on the top, which accommodates all three actions above the tabs, next to the screen title (the name of the listing). If you need more space on smaller devices, or if future design iterations add more functionality, some of the detail page actions can either migrate to the overflow menu or to the split action bar that's covered in the next chapter. Last but not least, remove all the containers on the screen, replacing them with Android 4.0 headers, following the Mobile Space Unbound principle discussed in Chapter 2. Note that frugal use of the real estate allows the redesigned screen to show several additional lines of text—no small feat on the mobile screen!

Bringing It All Together

Figure 1.20 shows three AutoTrader screens before the suggested redesign. Note the older IA and iOS treatment of the controls, fields, and buttons. Sections of the screen are separated from one another by containers with rounded corners. Within each section, the elements that implement interactivity are especially called out by the → symbol to separate them visually from noninteractive elements, giving the overall visual style a heavy appearance.

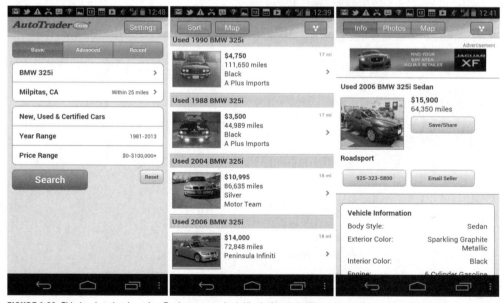

FIGURE 1.20: This is what the three AutoTrader screens look like before redesign.

In contrast, Figure 1.21 shows the three redesigned screens imbued with Android 4.0 DNA.

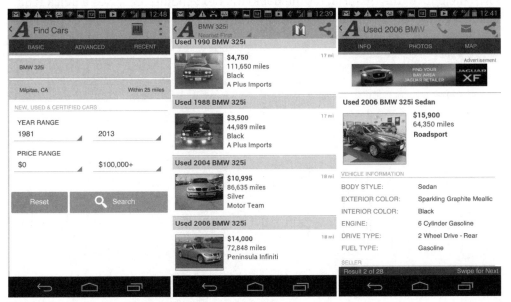

FIGURE 1.21: The three AutoTrader screens redesigned for Android 4.0.

In the redesign, a set of specialized touch controls and a unique navigational scheme recommended by the Android 4.0 guidelines are used. From a visual perspective, the new design uses flat buttons, touch panels, and action bars that are mostly devoid of gradients and rounded corners. Finally, the new design removes any containers and extraneous touch indicators.

Throughout the process you examined several different versions of each screen. This is natural: Android design is not complicated, but very sophisticated: with crushing space constraints and exciting novel interaction opportunities. That is why it's a great idea to customer-test new design ideas thoroughly before they are implemented, using quick, inexpensive prototypes. I prefer to do prototyping and testing with sticky notes, and throughout this book you see many examples of using this design methodology. Chapter 4, "Mobile Design Process," provides a detailed description of the entire design and prototyping process and offers practical techniques to tackle customer testing with confidence.

AutoTrader offered a great opportunity for showing the detail of the Android visual design language and widgets, which kicks off the book with a powerful example.

However, this is just a short overview of many innovative, interesting, and useful design patterns found in Android. Before digging into the design patterns that form the bulk of the material for the book, the next chapter provides a quick look at a few aspects of Android that make it different from the other mobile operating systems.

CHAPTER 2

What Makes Android Different

For many years since its release, the Android OS has been behaving like a teenager in the grip of raging hormones. Growth has been nothing short of explosive, and the changes have been sweeping and profound. With the release of Ice Cream Sandwich, the user interface (UI) standards and design elements have changed dramatically, and the platform has matured and stabilized somewhat. Nevertheless the OS has retained its rebellious hacker DNA with unique features that are authentically Android.

Welcome to Flatland

The first thing you might notice when comparing the Android OS apps with Apple iOS is that the world of Android apps is flat. Flat are the buttons. Flat are the content areas. And flat are all the toolbars and controls. Just like the Flatland people from Rudy Rucker's story, "Message Found in a Copy of Flatland," Android does not "see" anything outside two dimensions. Nor does it pretend to be anything other than a pure digital artifact: a thing imagined and created, not real in any physical sense. It's a piece of software that runs the hardware, not the other way around. And that, as far as I am concerned, is a very good thing. Why? Because dispensing with the need to make things "real" and "pretty" allows the content to shine and sets a stage for the authentic minimalist digital experience for your customers. In many ways, Android 4.0 uses a flat digital visual scheme similar to that used in Windows Modern UI, another mobile operating system that stands in sharp contrast to Apple iOS.

Compare, for example, the Android Messaging app with the iOS counterpart in Figure 2.1.

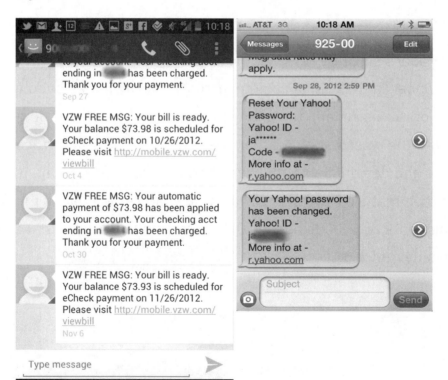

FIGURE 2.1: Compare the two Messaging apps from iOS and Android 4.0.

The first thing to notice is the information density: There is a great deal more content crammed on screen in the Android app. Part of the reason is that the iOS uses the "speech bubble" representation of the message, whereas the Android app is simply listing messages in the table. Boring? For some folks, perhaps. However, Android makes no excuses and no decorations—the app is a straightforward, flat, and highly functional SMS machine. The overall visual scheme is reserved, almost corporate. Notice also the toolbars: iOS dictates that the toolbar elements have a three-dimensional quality that makes the elements seem to pop off the page. This is achieved with the gradients that help digital objects like toolbars visually approximate the physical world. In contrast, the Android toolbars, and indeed the entire page is decidedly two-dimensional and completely free from having to look like a physical object. Unabashed embrace of Flatland and "freedom from three-dimensions" opens the door to creating menus that are semitransparent (see Figure 2.2) and commit fully to the "content-first" directive. (Various menu styles are covered in more detail in Chapter 13, "Navigation.")

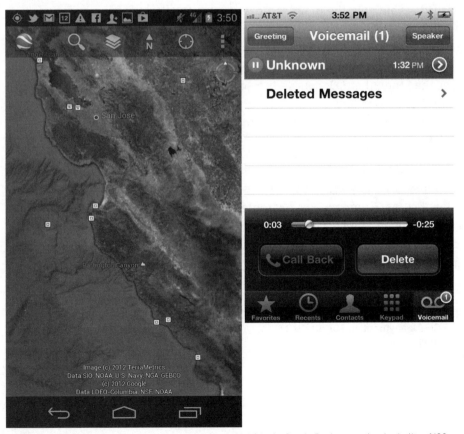

FIGURE 2.2: Compare the semitransparent menus of Android 4.0 in the Google Earth app to the physicality of iOS menus in the iPhone app.

The entire Android screen is built with grayscale, using just enough color to make the toolbars a bit darker, whereas the content areas mostly remain light. Color is one area in which many other mobile OSs, such as Windows Modern UI, contrast strongly with the Android Ice Cream Sandwich OS. Even though both design languages embrace the Flatland, Windows Modern UI veritably explodes with both color and interactivity, and the home screen literally "pops" with movement, as each element vibrates, flips, and slides with clever transitions and contrasting color combinations. In contrast, a typical Android screen looks compact, serious, business-like, and provides only the essentials—exactly like a typical wireframe. Even more interesting, this "flat world" scenario applies to buttons and tap-targets. Android "buttons" also have no gradients, which is the subject of the next section.

Tap Anywhere

In the early days of using the mainframe workstations (that was a *very* long time ago), I remember being stumped when first seeing the message Tap Any Key to Continue. Programming was such a precise discipline, that it seemed inconceivable that any key would work. You could tap *any* key, but which is the *best* key? Is one better than the other? Of course the confusion quickly passed, and I learned to enjoy the freedom to just jam anywhere on the keyboard without having to think hard—most of the time just hitting the space bar with my thumb.

Android takes buttons to a new level. Whereas the iOS painstakingly identifies any tap-worthy element with the three-dimensional beveled button, Android simply assumes that any element on the screen is a tap target, often providing no additional clues. Compare, for example, the two message rows in the messaging app shown in Figure 2.3. The iOS implementation provides the beveled, round circular → button, whereas Android in contrast provides...absolutely nothing.

VZW FREE MSG: Your bill is ready. Your balance $73.98 is scheduled for eCheck payment on 10/26/2012. Please visit http://mobile.vzw.com/viewbill
Oct 4

Your Yahoo! password has been changed. Yahoo! ID - ja█████ More info at - r.yahoo.com

FIGURE 2.3: Compare table rows in Android 4.0 and iOS.

In Android 4.0 the customer must figure out that she can simply tap anywhere on the element (that is, anywhere in the row) to get more information. The initial cognitive friction for taking the action is often high, especially if the customer is transitioning from iOS or previous Android versions. However, most people figure out the new visual scheme quickly: If a customer wants more information about something, she is (usually) able to tap it (even if an element has no visual tap affordance). Android trains customers to simply "tap any key to continue."

The Tap Anywhere visual design principle finds its ultimate expression in the typical Android buttons, which are implemented instead as "tap-worthy areas" (to borrow a favorite catch phrase from the mobile design expert and book author, Josh Clark, http://globalmoxie.com). In Figure 2.4, the iOS takes the time to make sure the buttons look and feel tappable with an elaborate combination of a prominent bevel, inner drop-shadow, and gradient.

FIGURE 2.4: Compare the Android tap-worthy areas to the iOS beveled buttons.

In contrast, Android commits whole-heartedly to the Tap Anywhere approach, using areas instead of buttons with just a hint of a vertical separator. This underscores the Android theme of not defining rigid visual border areas for tap targets and not wrapping touch targets with whitespace. This theme is profound because in its purest expression this means that *everything on the touch screen is a touch target*. For designers, this is both a challenge and an opportunity. It's a challenge because without primary or secondary tap targets, a consumer might be justifiably confused about the tap-any-button scenario: "If you can tap anywhere, which area is the best one?" The Tap Anywhere scenario also presents a hefty challenge for designers and developers: Everything that the customer would conceivably *want* to touch should be responsive and do something intuitive. This book discusses how to adhere to this design principle without losing sight of your customers' primary goals (and your development budget) in Chapters 10, "Data Entry," and 11, "Forms."

Tap Anywhere is also a tremendous under-exploited opportunity to introduce accelerometer gestures, multitouch gestures, and "hidden" menus that promulgate the content-forward design and promote immersive digital experiences. The myriad exciting possibilities are covered further in lucky Chapter 13, "Navigation."

Right-Size for Every Device

In the early days of Android, it became clear that simply porting the Apple menu model to every device would result in failure. Part of the reason was the tremendous range of different devices that were running Android: From the tiny HTC Hero, to the 7- and 10-inch tablets, to Android-enabled ski-goggles, smart homes, and in-car touch control panels—Android interfaces offer a great variety of space constraints in which customers must operate. Recall from the AutoTrader app case study in Chapter 1, "Design for Android: A Case Study," that simply hiding every function in the navigation bar menu on the bottom of the screen as Android had done for versions 2.3 and earlier was not the answer, either, because it hid vital functions for devices that actually had the space to show them, as well as made having more than three or four contextual actions difficult (see Figure 2.5).

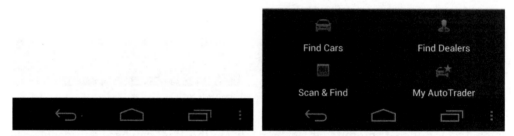

FIGURE 2.5: This example from the AutoTrader app demonstrates how the hardware menu functions for Android 2.3 and older hide essential functions.

To solve this problem, Android 4.0 designers used an authentic mobile solution: the overflow menu. To understand how to design for Android 4.0 and 4.1, it is essential to understand how this menu works. Functions are distributed to one or more menus called action bars. When the interface distributes functions along more than one action bar, the second action bar is called...get ready for it...a split action bar. An app with two action bars is shown in Figure 2.6.

Regardless of the number of bars (typically one or two: the action bar on the top and the split action bar on the bottom), the menu acts as an accordion, expanding and contracting with the available screen real estate. Smaller screens get only a few essential functions. Larger screens such as in tablets get the entire available menu, with a bias toward having only a single action bar. Additional functions that do not fit on the action bars end up in the overflow menu, which is an excellent, highly scalable, mobile solution to the problem of limited real estate. Figure 2.7 shows a comparison of menus on a tablet (top) and a phone (bottom).

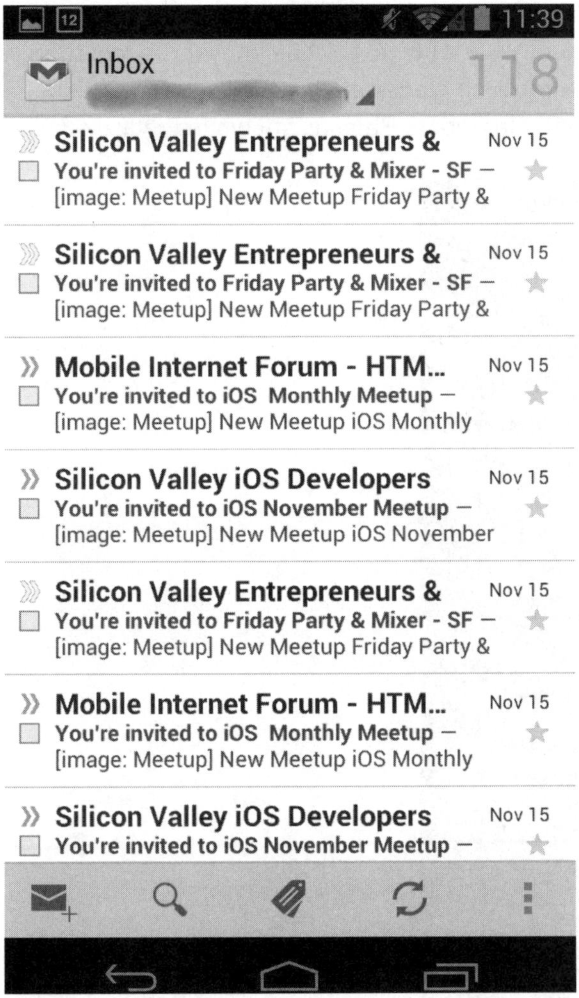

Inbox

118

>> **Silicon Valley Entrepreneurs &**   Nov 15
☐ You're invited to Friday Party & Mixer - SF —   ⭐
   [image: Meetup] New Meetup Friday Party &

>> **Silicon Valley Entrepreneurs &**   Nov 15
☐ You're invited to Friday Party & Mixer - SF —   ⭐
   [image: Meetup] New Meetup Friday Party &

>> **Mobile Internet Forum - HTM...**   Nov 15
☐ You're invited to iOS Monthly Meetup —   ⭐
   [image: Meetup] New Meetup iOS Monthly

>> **Silicon Valley iOS Developers**   Nov 15
☐ You're invited to iOS November Meetup —   ⭐
   [image: Meetup] New Meetup iOS November

>> **Silicon Valley Entrepreneurs &**   Nov 15
☐ You're invited to Friday Party & Mixer - SF —   ⭐
   [image: Meetup] New Meetup Friday Party &

>> **Mobile Internet Forum - HTM...**   Nov 15
☐ You're invited to iOS Monthly Meetup —   ⭐
   [image: Meetup] New Meetup iOS Monthly

>> **Silicon Valley iOS Developers**   Nov 15
☐ You're invited to iOS November Meetup —   ⭐

FIGURE 2.6: The Gmail app has an action bar at the top and a split action bar at the bottom.

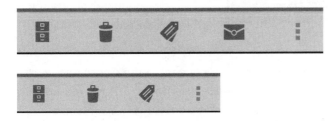

FIGURE 2.7: In the Gmail app on smaller screens, extra functions move to the overflow menu.

In stark contrast to iOS and older Android menus, the action bars, as a rule, use only icons, whereas the overflow menus use only text. Icons and words together are still used in Android 4.0—for example, in places such as some Cancel/OK action buttons and in the Drawer menu in Google Plus, as shown in Figure 2.8.

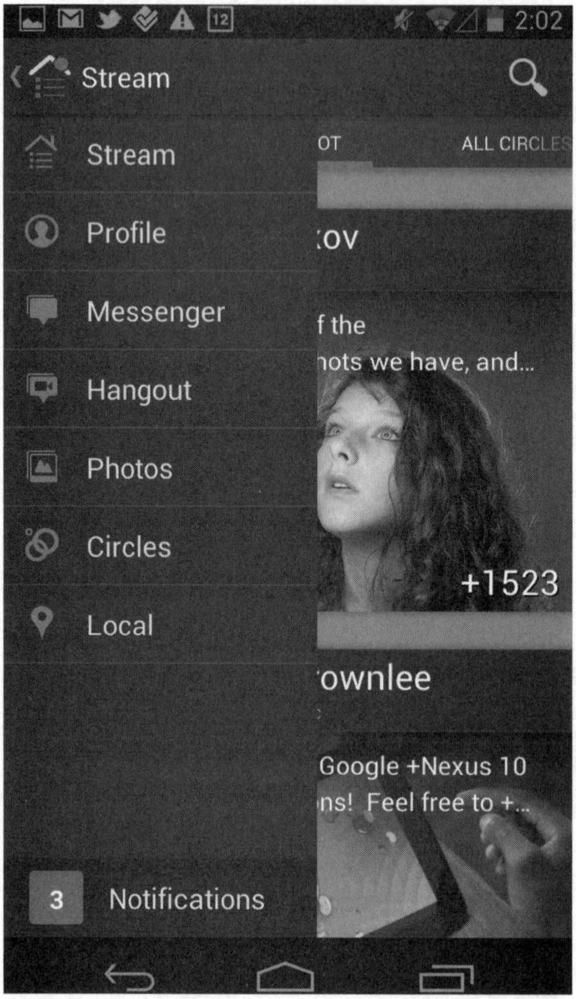

FIGURE 2.8: The Google Plus app uses icons and text in the Drawer menu.

Action bars combined with the overflow menu usually enable the Android to work reasonably well on a majority of screens and device orientations. However, not all resulting UIs are ergonomically desirable: Specific device constraints are discussed in the next chapter. Chapter 13 discusses how to make the most of this pattern using the "hidden" menus and active corners in Swiss-Army-Knife Navigation, and Chapter 8, "Sorting and Filtering," shows you different menu approaches for essential functions like Search. Discover how to design effective 7- and 10-inch tablet UIs in Chapter 14, "Tablet Patterns."

Mobile Space, Unbound

One important corollary to the principles of Tap Anywhere and Right-Size for Every Device is the unabashed removal from the interface of containers of any kind, in stark contrast to iOS, which typically features multiple containers with rounded corners, as shown in Figure 2.9.

FIGURE 2.9: The iOS Settings app demonstrates emphasis in iOS on UI containers.

Many designers point out that the defining features of the Windows Modern UI (see Figure 2.10) are the multicolored square containers, called *tiles*, as well as the unifying backgrounds and sweeping headers in the Panorama controls (read more about those in Chapter 13).

FIGURE 2.10: Windows Modern UI Settings app demonstrates the sweeping backgrounds and titles in a Panorama control.

In contrast to both iOS and Windows Modern UI, one of the signature features of Android 4.0 are the simple headers that completely dispense with containers. Instead of containers, when separation of the UI elements' sections is necessary, Android 4.0 OS uses the contrasting colored headers, executed in uppercase Roboto font and underlined with a horizontal divider, as shown in Figure 2.11.

This general absence of containers promotes flow of the forms and uses every pixel to emphasize the content. The absence of containers also helps the UI look great on a variety of devices, regardless of the width and height of the screen.

Containers behave best when they are smaller than the size of the screen; in other words, the screen shows more than one container. For this reason, containers in iOS often behave awkwardly when the device is positioned in the horizontal orientation. This is a special consideration for Android because, as discussed earlier, Right-Size for Every Device means that the interface must work for every screen, no matter how small or strangely proportioned (within reason of course). Thus containers as an interface principle do not work well because they cannot be adjusted dynamically for every situation. Instead, the content flows free, unbound, and adjusts appropriately to the size of the screen.

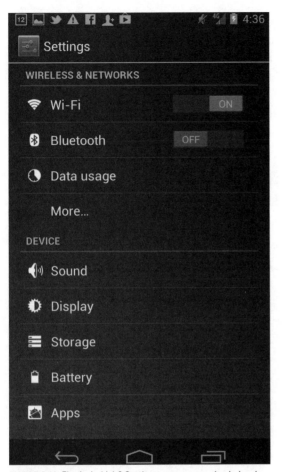

FIGURE 2.11: The Android 4.0 Settings screen uses simple headers instead of containers to promote page flow.

Unfortunately, nothing on the mobile platform comes free, and the absence of containers is no exception. Although form flow is enhanced, the absence of containers places an extra burden on vertical spacing to help separate the form fields from one another. On smaller devices sometimes this can be difficult, resulting in the forms that simply seem to go on forever without any idea where the person is at any given moment. The other issue is the minimalist color treatment. In the Settings screen, the headers are rendered in light gray, which is similar to the color of the links. In other native apps like Calendar, the headers are colored in the same way as the active fields; Both are light blue. Either treatment can cause confusion because it is not clear which are the active tappable fields and which are the headers. Using a header color that is visually distinct from both the active links and active fields is a good basic usability practice and usually helps resolve the confusion.

Think Globally, Act Locally

One of the most interesting and brave principles of the new Android 4.0/4.1 design scheme is the local-actions-first principle. Practically all other OSes on the market, including the older Android OS versions, take special care to make global navigation available throughout the app. For instance, Apple iOS does this through the prevalence of the tab bar, as shown in Figure 2.12.

FIGURE 2.12: The Amazon.com app uses the iOS tab bar to make global navigation accessible throughout the app.

The same principle used to apply to the older, pre-Ice Cream Sandwich Android apps, such as the pre-4.0 Android Amazon.com app design, as shown in Figure 2.13.

In contrast, Android 4.0 offers the customers a brave, new world in which the actions presented to the customer are always the most appropriate ones to the task at hand.

For example, in the Mail app as shown in Figure 2.14, the key actions in the action bars on the app's summary pane are Search, New Message, and so on, plus a few more general actions such as Settings, Help, and Send Feedback, which are hidden in the overflow menu.

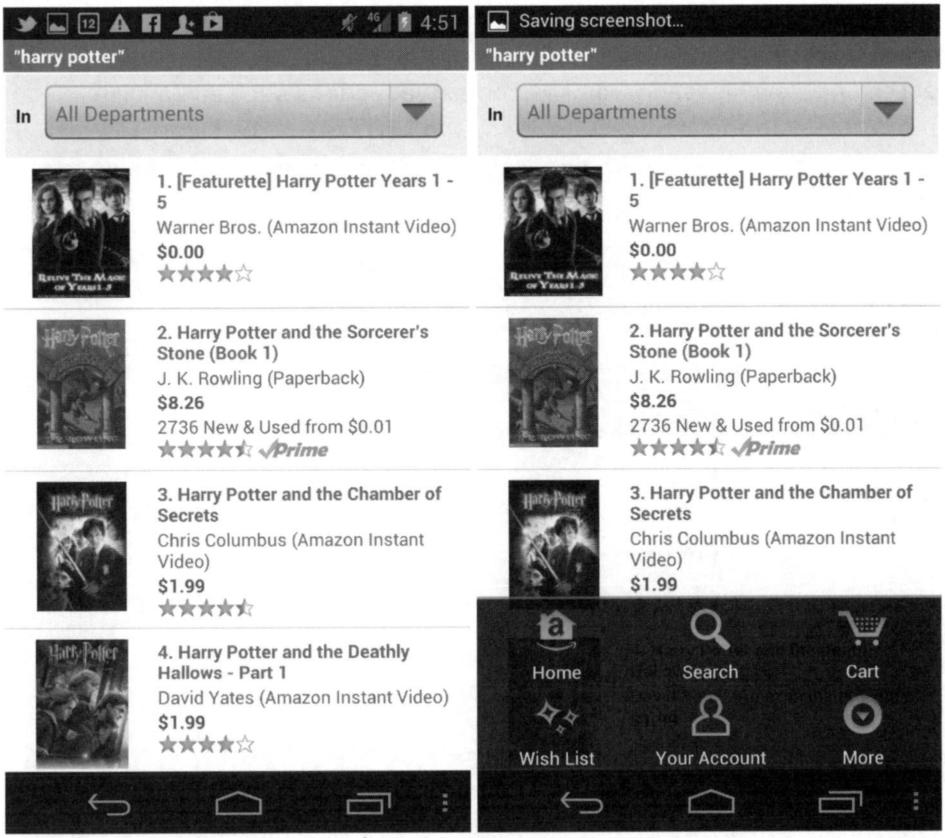

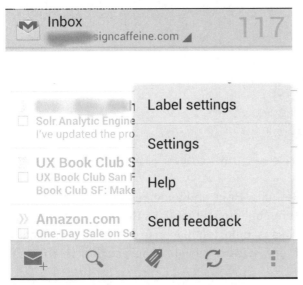

FIGURE 2.13: Older Android apps have a global menu in the navigation bar as in the Amazon.com app.

FIGURE 2.14: The Gmail app's List view shows global actions.

Drilling down into the specific piece of mail, however, makes a dramatic change to the available actions. In Figure 2.15 instead of the top-tier menu items, such as Search and New Message, you see contextual actions such as Favorite and Reply (the star and arrow icons on the top bar), as well as commonly used Archive, Trash, and Tag (the file cabinet, trash can, and tag icons on the bottom bar).

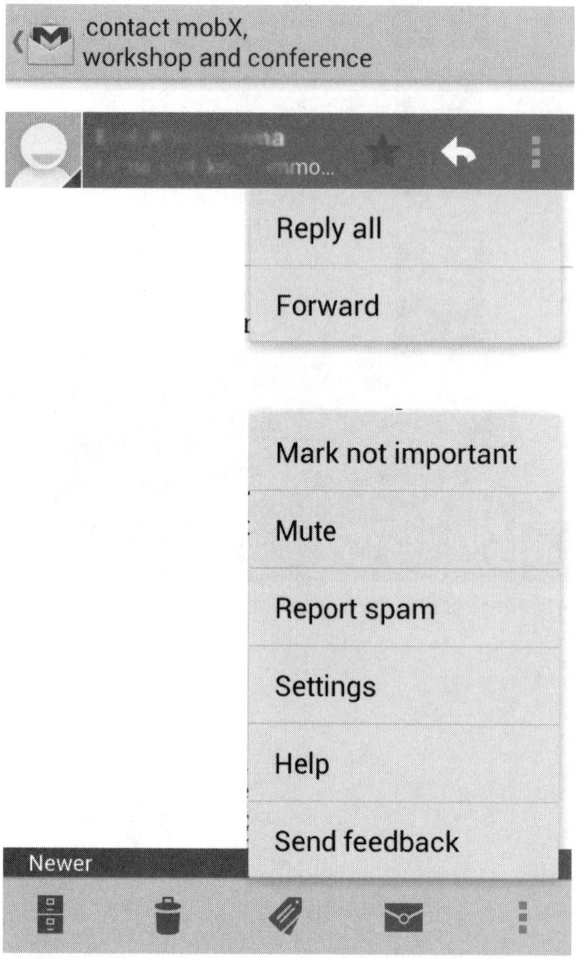

FIGURE 2.15: The Gmail app's detail view shows contextual actions instead of the global menu.

The more general action such as Settings, Help, and Send Feedback are still available in the bottom overflow menu, but global functions that were available in the list view are completely gone. This is important because due to the Act Locally principle, from the e-mail detail screen, it is impossible to access Search and

New Message, for example. The Act Locally principle also extends to the overflow menu: The top overflow menu is not merely an extension of the message functions but instead augments the Reply options with less frequently used functions, such as Reply All and Forward.

The Android 4.0 screen label is yet another corollary to this design principle. For example, in the iOS, the screen label tells the customers where they are and the back button often spells out where people will be going should they tap it. As in the mail app example shown in Figure 2.16, iOS Mail app uses 14 of 69 as a screen title, and the Back button displays All Mail (41).

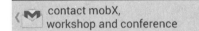

FIGURE 2.16: Compare the Android title header on the left with the back button and screen title in iOS on the right.

Android 4.0 offers no indication of where customers would go should they tap the back button. Instead, the entire bar is devoted to a screen label showing the subject of the e-mail—that is, *where the customer is at the moment*. This creates some confusion, especially among the iOS designers and customers taking on the Android design patterns. To these folks, "← + logo + label" means that the action would be *going back to* the destination shown on the label, in sharp contrast to the Android 4.0 where the same visual treatment means "you are at the label location" and "tapping the back button would go up a level." Unfortunately, there is no good solution at the moment to this mental model transition, other than simply perhaps to get used to it.

If you contrast this hyper-local screen labeling with the older Android OS breadcrumb treatment, such as the one shown in Figure 2.17, you can see that the change is actually profound. The Android 4.0 OS simply ignores anything other than the hyper-local context.

Using only local functions on each of the screens is a departure from the earlier versions of the mobile technology—one that signals that Android assumes that people are no longer going to be "lost" without some global tab navigation showing them at all times where they are. Instead, Android 4.0 uses the "Think Globally, Act Locally" principle: The hyper-local actions and screen labels remove the global navigation (that used to be in the breadcrumb) from being in front of the customer. At the same time, in the back of their minds, Android 4.0 customers have to understand that to get to *all* the global navigation, they must press the back button one or more times.

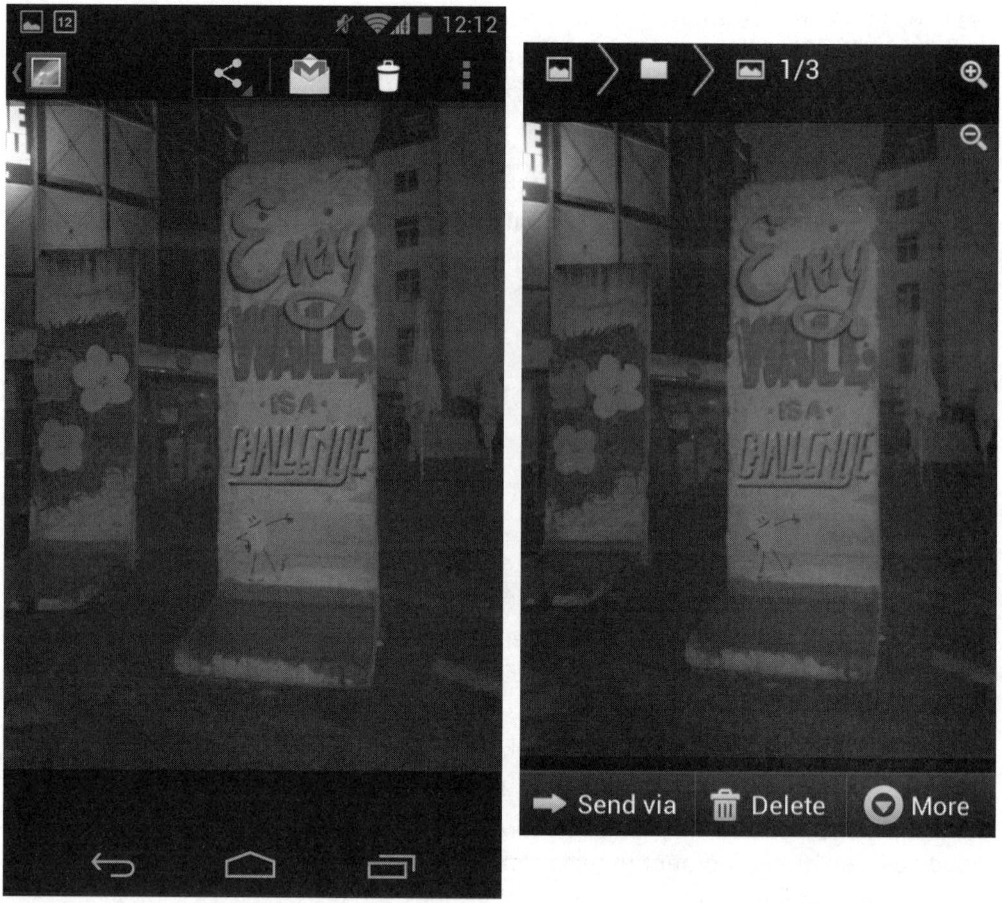

FIGURE 2.17: Compare the Android 4.0 action bar from the Photo Gallery app on the left to the Android 2.3 breadcrumb on the right.

Interestingly, this hyper-local context does not exactly fit with many apps' best use cases. For example, Facebook is a notable exception that uses the Swiss-Army-Knife top-left navigation to make global actions universally available. Some practical design patterns for addressing this tension between global and local are covered in Chapter 13.

Now that you've had an overview of the general design principles that make Android 4.0 different, it's time to dig into how these principles are implemented in a variety of devices that support Android.

CHAPTER 3

Android Fragmentation

Like a shrapnel grenade in the ball-bearings factory, Android fragmentation has now reached epic proportions. Here's how to sort out what's important and create the app User Experience (UX) design strategy that works for various device flavors.

What's Fragmentation?

According to OpenSignal, in April 2012, 6 months prior to the time this book was written, there were more than 3,997 distinct Android devices (based on a study of more than 681,900 devices: http://opensignal.com/reports/fragmentation.php). The number-one brand was Samsung, with 40-percent market penetration. HTC, SEMS, and Motorola *together* accounted for approximately 30 percent of the market share. Approximately 9 percent of the total devices were Samsung's GT-i9100 (the Galaxy SII), but beyond that there is little in the way of the trend, including multiple one-offs such as the Concorde Tab (a Hungarian 10.1-inch device), the Lemon P1 (a dual SIM Indian phone), and the Energy Tablet i724 (a Spanish tablet aimed at home entertainment). For those that don't mind the price tag, the famous Swiss watchmaker Tag Heuer released The Racer: A $3,600 Android smartphone with "unparalleled torsional strength" built with carbon fiber and titanium elements and a shockproof rubber chassis. (I don't know about you, but I think I'd rather fancy one—http://www.afterdawn.com/news/article.cfm/2012/03/13/tag_heuer_launches_carbon_fiber_android_phone_with_huge_price_tag).

Staggering diversity did not end with devices and brands. The study found that Android OS versions are likewise varied, with only approximately 9 percent of customers using Android 4.0 or above, and approximately 76 percent of customers using Android 2.2 and 2.3. In addition, most manufacturers further customize each Android OS version in hopes of achieving market differentiation.

Screen resolution was similarly varied, presenting an almost continuous range from insanely generous 1920 × 1200 and 2040 × 1152 to a tiny and barely usable 240 × 180. And this fragmentation is likely to get worse in the near future as an even greater variety of screens and devices, such as TVs, smart refrigerators, Android ski goggles, and 3-D screens and projectors, will be hitting the market shortly.

How is a designer to deal with this staggering complexity?

Everything Is in Time and Passes Away

One of the Buddhist Noble Truths is that everything that arises passes away. Nowhere is this more true than in Android development. It is worth remembering that even as recently as one year ago, HTC was a dominant Android device brand with the EVO lineup taking the lead. As of November 2012, the most popular device is now Galaxy SIII; the dominant Android version is now 4.0+, with 4.1 (Jelly Bean) starting to catch on quickly. Although Samsung is still a dominant brand, the Google Nexus line of phones and tablets made by ASUS are strong contenders, with LG not far behind—at least in the United States. In Germany, the smaller and cheaper Sony phones are reigning supreme. As of the time of this writing, Sony phones are virtually unknown in the United States.

The bottom line is that if your app works on today's top two or three models of the phones and tablets, you are generally in great shape. You don't need to worry about what was around 6 months ago or what's coming around the corner because it has not been invented yet! The point is not to sweat the small stuff. In other words, most of the phone models, versions, and manufacturers that are #1 today will be displaced from the top spot six months from now. Consider once-great mobile brands that today are barely remembered:

- Palm and the Web OS

- Motorola, which was swallowed by Google and quickly lost its hardware leadership

- Nokia, which was bought out by Microsoft and all but disappeared off the face of the Earth

- Blackberry OS, which is barely hanging on to a tiny fraction of market share with the skin of the right thumb

The important thing for design is not the latest gadget. It is a set of touch technology Android *device trends*, which do not change as quickly because they are based on the ergonomics of a human interacting with basic touch-screen technology. These device trends form the basic DNA of the mobile and tablet design patterns which changes slowly over time.

Android Device Trends

When flexible screen material is popularized, with touch panels on the back of the device and other interesting near-future technology advancements, the device trends described here will change, of course. For now, the thickness and weight of the device; the cost of components; the size of the screen; and the size and flexibility of hands, fingers, and pocket sizes of modern clothing dictate the basic range of sizes of mobile devices and interactions with them. Over time it seems that the variety of touch-screen Android devices has standardized itself around five basic sizes: compact phones, full-size mobile phones, tablet-phone hybrids, small tablets, and large tablets. Much more than various screen resolutions, these classes of devices present specific design decisions and ergonomic requirements for app interaction design, as discussed in the following sections.

Compact Phones

These are small, usually inexpensive Android devices. As of this writing, Kyocera Milano is a good example. These devices are tiny, 4.5" × 2.5", with a screen of only about 1.8" × 2". Because of this, the entire area of the device can be accessed easily with a thumb of one hand, as shown in the "hot zone" (the area most accessible when a customer uses the most common grip) on the right in Figure 3.1. Unfortunately, the tiny screen means that the hand holding the device also covers much of the bottom screen area, even while the customer is simply holding the device, which makes split action bar navigation problematic.

FIGURE 3.1: Compact Phone Kyocera Milano is used to demonstrate the right-hand grip hot zone for the touch screen.

Note that the buttons are forced to be taller, and there is only room across the screen width for three or four buttons. This is the issue with the smaller devices that need not only to show the on-screen text and icons but also present a suitably large touch target. In fact, simply having two navigation bars on the screen would take up more than 30 percent of the screen area!

There is room for only a single action bar on the top of the screen. This top action bar can display only two or three functions accurately (only one function if a screen title is used). Because of the restricted screen real estate, creating an immersive user interface with a semitransparent Swiss-Army-Knife navigation pattern, which uses zero screen real estate for menus (so the entire screen is devoted to content) makes a lot of sense. See Chapter 13, "Navigation," for some ideas on this topic. Be aware that the menu, when opened, will likely take over the entire screen, as will most secondary selection controls such as the picker wheel.

The on-screen keyboard is barely usable, so contrary to logic these tiny devices tend to come with a small, cheap, fold-out hardware keyboard, which barely improves the already poor input accuracy. Hardware keyboards are not that different from their on-screen counterparts. The major difference is that the screen does not enter the "extraction" mode (See the "10.7 Pattern: Free-Form Text Input and Extract" section) when the hardware keyboard is enabled. Instead, the orientation of the screen simply changes from vertical to horizontal, so the form is a bit wider. The input is then performed via the hardware keyboard, and the form navigation is done by scrolling the form using the touch screen. There is little difference between the hardware and software keyboard modes of input otherwise.

Full-Size Mobile Phones

This is the most popular size of devices on the market. At the time of this writing, a good example is the Samsung Galaxy SIII, 5.4" × 2.8", with a 4.8-inch diagonal screen size. The screen resolution is purposefully not mentioned. Actually, in doing mobile UX research, the way the device is used has little to do with screen resolution and more to do with the size, dimensions, and weight of the device. If the resolution is too low to display the required set of touch icons across the screen, then the behavior changes. However, for most modern devices the resolution is already adequate. Adding more pixels improves the picture and makes for great marketing campaigns that motivate people to buy a new gadget; it has little effect on the customers' behavior after they acquire it. The one distinguishing factor behind the mobile phones is that they tend to be light enough and small enough to be used *one-handed*. The hot zone for a full-size phone in a right-handed grip is shown in Figure 3.2.

FIGURE 3.2: This is the hot zone for a right-hand grip on a full-size mobile phone, Samsung Galaxy SIII.

As the right-hand hot zone in Figure 3.2 shows, the bottom action bar is easily accessible, *however the ergonomics of accessing the top action bar don't quite work out*: Most customers need to use their left hands to tap the controls on the top of the device or have to reposition the device and stretch their fingers awkwardly. This is especially true for women and teens, or customers with smaller hands. Another drawback of commands on top of the screen is you have to cover the screen to reach them. This is not exactly toeing the Android "Party Line" because, unfortunately, the top action bar is where the Android guidelines recommend placing most of the key functions. Indeed, that's where the key functions reside in the majority of the Google Android apps as of the date of this writing.

In the United States, the Android phones tend to come in larger, literally pocket-bursting sizes. This is partly due to the current market penetration of the Apple iOS iPhone. It helps marketing campaigns define differentiation if the Android devices have a visibly larger screen, so the full-size Android mobile phones tend to be larger than the iPhone's 3.5-inch screen. Although this is a strong trend, larger size is not always the case. As already mentioned, in Europe slightly smaller and less expensive Android phones made by Sony are about the size of the iPhone 4. Anything smaller than the size of the iPhone 4 can safely be considered to be a compact phone and treated accordingly.

On these smaller Android phones that have the 3.5-inch screens, it is much easier to reach the top action bar; although it still requires awkward juggling. The result is that features such as the Drawer navigation in the Google Plus and Facebook apps are actually hard to reach. As discussed in Chapter 1, "Design for Android: A Case Study," one workaround is to use the left-to-right swipe gesture to bring out the navigation drawer, as indicated by the bevel on the left side of the screen. Chapter 13 discusses using bottom corners for immersive Swiss-Army-Knife navigation that are more easily accessible than the top corners. Keep this in mind as you design your own apps. Also consider using gestures like the C-Swipe that's discussed in Chapter 14, "Tablet Patterns." As this book is being written, these gestures are not part of the Android 4.0 guidelines, but they do work in the real world.

Until recently, a differentiating factor for many Android phones was a slide-out hardware keyboard. This was mainly a marketing push designed to entice would-be iPhone users who were skeptical about the usability of on-screen typing as well as current Blackberry users unwilling to part with their hardware keyboards. In the current device lineup, however, the keyboards are generally absent, except for a handful of models from manufacturers such as Motorola. This trend reflects the general acceptance of the soft keyboards by the market, but also the larger screen sizes that are more forgiving of fat fingers, as well as software-driven keyboard input improvements such as predictive text, Swype, ShapeWriter, SlideIT, and so on. You can find more information about these in the Android Forums at http://androidforums.com/android-applications/50081-swype-vs-slideit-vs-shapewriter.html.

Tablet-Phone Hybrids

Tablet-phone hybrids such as the Galaxy Note are a conundrum. For most apps, visibly, there is no improvement in the screen resolution, so the person using the awkward device gets the same on-screen experience as the full-size mobile phone customer. However, bringing one of these giant hybrid phones (roughly one-inch taller and one-half an inch wider than the big full-size mobile phones) to one's ear for making a phone call is rather awkward, to put it mildly. Yet as the success of the Galaxy Note has proven, people are actually rather interested in these devices. Part of the appeal is the larger screen that enables comfortable e-book reading and mobile web surfing.

Hybrid phone customers find that for the vast majority of tasks, the use of the device is a dedicated two-handed affair. Although it's possible to easily hold Galaxy Note in one hand, even if you have Niccolò Paganini's legendary long fingers, it is almost impossible to use that same hand to reach and operate the top action bar functions. As shown in the hot zone in the Figure 3.3, one hand is not enough to reach all the functions of the tablet-phone hybrid, which forces an asymmetric one-handed grip with one hand holding the device while the other hand does the tapping.

FIGURE 3.3: The Samsung Galaxy Note's limited one-handed hot zone forces the customer to use a two-handed grip.

Most of the apps on the device are not customized to take full advantage of the larger screen size; some apps, most notably the Calendar, offer additional features such as slide-out tabs, which are interesting interface approaches but can still be rather awkward to use. My recommendation for these devices is *not* to customize the native app experience unless a large percentage of the app's customers are using one of these hybrids, or you are designing a native app specifically customized for this device class.

Gesture-based menu interactions, like the slide-out drawers and C-Swipe mentioned earlier, would actually enable one-handed use because these hybrid devices are light enough and compact enough to be held comfortably in one hand. This is because the device's *center of gravity* still falls comfortably within the reach of the fingers of one hand. Most of the issues of one-handed use come from the inability to *reach* the navigation components, not the device size or weight distribution, as is the case for true tablets.

Small Tablets

Small tablets, such as the 7-inch Samsung Galaxy Tab 2, offer some unique use cases and challenges. Whereas all the previous devices were mostly used in a vertical (portrait) orientation, tablets are more often rotated, doubling the complexity of the interface (see Figure 3.4).

FIGURE 3.4: The 7-inch Samsung Galaxy Tab 2 is an example of a small tablet device.

As the hot zone picture in Figure 3.5 shows, in a vertical orientation, most small tablets are held with one hand, with the second hand tapping the controls. On the other hand (literally) in the horizontal (landscape) orientation, the device is most often held in a committed two-handed grip.

FIGURE 3.5: Small tablets dictate different hot zones for one-handed vertical, two-handed vertical, and two handed horizontal orientations.

Why this distinction? Holding the device in horizontal orientation in one hand is simply more difficult than holding the same device vertically in one hand. (Really, try it!) It has to do with a center of gravity of the device and where the center of gravity is located with respect to the fingers and the wrist. This is a simple physiological reality and is not likely to change any time soon: The fundamentals such as device weight, dimensions, and materials must change first.

What does all this mean from the standpoint of the interface design? There is enough space on the small tablet device to show one or more action bars comfortably. Looking at the hot zone in Figure 3.5, on small tablets in vertical orientation, the entire top and bottom action bars are easily accessible, with the top bar being slightly more so. Thus Android guidelines can be followed as-is, and there is no need for measures like immersive navigation. Swiss-Army-Knife navigation measures can lead to *less* satisfying experiences on tablets because they hide essential functions and force customers to learn where they are, which can cause cognitive friction. This guideline does not apply to immersive tasks such as reading and gaming, which traditionally use lights-out navigation, but to more navigation-centric tasks such as shopping.

One size definitely does not fit all! In general, it's a good idea to make an effort to avoid cognitive friction for tablets, where the best interfaces keep customers in a state of flow, offering simple, intuitive functions and wide, sweeping gestures (such as swiping to turn the page) that come naturally with the larger touch device. The C-Swipe gesture can be used for either the hand that is holding the device, or the hand that does the tapping, but chances are that the C-Swipe will happen in the middle of the device, or near the top portion of the screen, not on the bottom as it would for a one-handed grip on the mobile phones.

Things might be a little different in the horizontal orientation, which calls for a committed two-handed grip. Most of the navigation functionality is accessed using thumbs, while the rest of the fingers rest on the back of the device. Fortunately, most of the screen surface area, (including the essential Android back button) is easily accessible with average thumbs on a small tablet with a two-handed grip. Thus again the Android guidelines can be used out-of-the-box; actually, they seem to be developed especially for small tablets!

Which is the preferred orientation? This question is difficult to answer, and most often no clear distinction can be drawn, even for specific tasks. From doing research and hours of field observations, people tend to read mostly with the device in a vertical orientation, where the interface forms a single column. On the other hand, the keyboard in vertical orientation does not offer the greatest experience—most people tend to rotate to horizontal when needing to use the keyboard-centered tasks such as filling out a form. Unfortunately, it is hard to see much of the form on a 7-inch tablet while the soft keyboard is also shown on the screen, so the entire form interaction on 7-inch tablets tends to be a rather awkward shuffling from vertical for scrolling to horizontal for typing, which is a less than ideal situation. This problem and some creative solutions are covered in detail in Chapters 10 ("Data Entry") and 11 ("Forms").

Another unique quality of small tablets is that they are actually small enough and light enough to be both held and manipulated with one hand, albeit with limited capability. Where you see this show up is in longer-term immersive activities such as reading. The person may want to both hold the small tablet and flip pages without using the second hand. This works well for the *right-handed* grip near the middle of the tablet, especially if the tablet is also partially resting on the person's lap, table, or chair arm. This is in part due to the generous margin around the touch screen, which enables the grip to be solid without touching the screen. Unfortunately, the same is not the case for the left-handed grip; most book apps flip the pages *backward* when they are tapped on the left side of the screen. The solution is to put controls along the left and right side of the screen or use a C-Swipe gesture to call up the menu from anywhere, as discussed in Chapter 14.

From what can be observed in the field, the entire one-handed grip-and-use approach is not the most solid use case for the Apple iPad mini. The margin around the screen is simply too small to enable a comfortable grip.

Finally, what actually defines the tablet as "small"? It is the quality of comfortable accessibility of the entire screen by the target customer population, without letting go of the sides of the device, with a possible one-handed vertical grip-and-use hold that defines the tablet as "small." When some of the screen becomes inaccessible or a tablet can no longer be held and used by the same hand, it can be classified as "large."

Large Tablets

You might think that large tablets such as the 10-inch Samsung Galaxy Tab 2 (see Figure 3.6) have been around long enough for people to get used to the unique interface challenges these devices present. However, in the Android 4.0, much of the interface guidelines are not adopted well for large tablets. Instead they are treated by Android exactly like their small counterparts. However, from the UX-perspective, although there are some similarities, there are also important and significant differences, as I point out here.

FIGURE 3.6: Large tablets such as the 10-inch Samsung Galaxy Tab 2 offer unique design challenges.

First, the soft keyboard works well in both orientations, which is especially noticeable when keyboard use on large tablets is compared with that of small tablets in the vertical orientation. On large tablets the vertical keyboard is more usable than it is in smaller tablets. However, there is also much less in terms of vertical space constraint on a 10-inch device when you're using the horizontal keyboard, so horizontal tends to be the preferred orientation for apps that require data entry, web browsers, and many other applications. Anecdotal observation suggests that reading tasks tend to be split about 50/50 between horizontal and vertical orientations, based on the reading medium and personal preference.

Unlike small tablets, large tablets do not allow one-handed unsupported grip-and-use hold. There is no balancing a larger tablet in one hand; instead, the device in either orientation is held in a committed two-handed grip that supports the heavier, more awkward (and more expensive) device securely. Figure 3.7 demonstrates the hot zones for a large tablet in a two-handed grip in vertical and horizontal orientations.

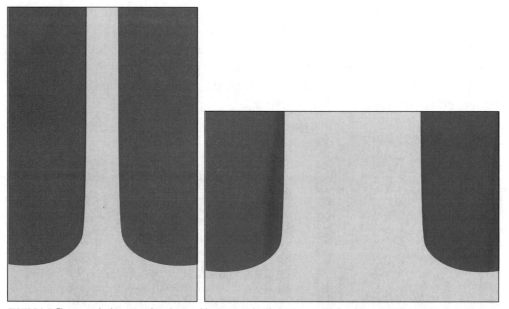

FIGURE 3.7: These are the hot zones for a large tablet in a two-handed grip in vertical and horizontal orientations.

For larger tablets in either orientation, the bottom of the device is not nearly as accessible as the top. That is because the bottom of the tablet is usually resting on some surface, like the person's lap or a table. This makes accessing controls on the bottom of the screen (including the essential back button!) awkward at best. The same goes for the controls located in the middle of the top action bar. Your customers will have to let go of the device to tap there, in which case Fitts's law takes over: The time required to rapidly move to a target area is a function of the distance to the target and the size of the target. If the target is small, such as tapping a button or an icon on the top action bar, it makes unsupported free-hand movement required to reach the target quite awkward, especially if it has to be repeated over and over. Mobile designer Josh Clark brilliantly calls this an "iPad elbow" (which you can rename with a slightly more utilitarian "large tablet elbow"). Large tablet elbow is especially pronounced for repeated tasks in a horizontal orientation, where reaching the middle on the top action bar or any portion of the bottom action bar is particularly awkward.

What does this mean from the interface perspective? As discussed in great detail in Chapter 14, one idea is to use only a small portion of the vertical action bar and forgo the bottom bar. If you're more adventurous you might want to try navigation and functional controls that run vertically along the left and right sides of the large tablet, instead of horizontally across top and bottom. Another idea is to use the C-Swipe to call up the menu in any part of the screen, an approach also discussed in Chapter 14. Neither of these approaches fits within the general Android guidelines and will definitely go against the "party doctrine." However, for the adventurous product team, native tablets apps offer the best opportunity for experimentation and creative differentiation.

Celebrate Fragmentation

Fragmentation is always going to be a challenge in Android app development. Yet as the OpenSignal article points out, there is much to be celebrated. Android has reached more than 195 countries, which is more places than most people visit in a single lifetime. Even more impressive is that the five countries with top Android use are the United States, Brazil, China, Russia, and Mexico, which means the developing world is leading Europe, at least in Android mobile computing. With cost of the Android hardware further dropping on a daily basis, your app is just as likely to be downloaded by a rural farmer in India as by a New York stock broker. Or if you look at it another way, your work can reach billions of people around the globe and make a real difference in their daily lives.

The mobile design is all about the context. To ensure that your app solves the right problems for the right audience, you need to customer-test your app as early and as often as possible with your target customers. Read the next chapter to discover a cheap and effective way to do just that using sticky note prototyping methodology.

CHAPTER 4

Mobile Design Process

What makes an effective mobile design process? At the end of the chapter, I describe an end-to-end mobile design process case study that showcases the sticky-note mobile design methodology used throughout the book. However, before you jump to this section, I want to discuss the challenges of designing in the mobile age and some approaches to adopting the classic User-Centered Design (UCD) techniques to the new medium so they remain effective and relevant.

Observe Human-Mobile Interaction in the Real World

In the past, context was one of the considerations when designing software, but it often took the back seat to other methods of analysis. Why is that? Because, ladies and gentlemen, before the arrival of mobile devices, the context was a computer that the customer would sit in front of, unless you were designing a computerized coffee maker (vile abomination!). Thus *at the moment of interaction with your software*, your customers were basically sitting down in chairs in front of computer screens with keyboards and mice.

In contrast, *in mobile design, context is king*—context that you and your team should ideally observe first hand, in a real world. It is no longer possible to reliably imagine and model how the interaction would proceed (a person sits in front of the computer, grabs the mouse, and so on) because the person's behavior and her interaction with the device is highly dependent on context. Even fundamental design parameters like device orientation and hand grip change dramatically when the customer is standing on a busy street corner and looking at a map; sharing photos of the kids while sitting on the couch with a spouse; talking with one's boss one-handed while trying to park the car; or reading while riding the city bus. To really understand what happens, *you and your team have to get out there to observe these interactions first hand, as they happen*. And while you are out there, simply asking questions is no longer enough to get accurate and precise data. To make solid design decisions, ideally you should observe *behaviors* using a realistic prototype of your app as a tool for eliciting these behavioral responses.

Your Prototyping Methods Must Allow for Variety in Form Factors

For many years, the same old fight between Mac and PC has dominated the tech landscape, followed by the browser wars. From the standpoint of the User Experience (UX) design, the PC was somewhat different from the Mac, but from the standpoint of the customer, the two were perhaps not *that* different—both operating systems used a mouse, keyboard, and large screen. Also, because the majority of software being produced was built for the Internet browser, the experience was largely device-independent. Yahoo! and Facebook looked similar in Internet Explorer on Vista and in Safari on Mac OS X.

In contrast, the age of mobile touch computing yielded a tremendous variety of platforms and device form factors. Small phones, large phones, small tablets, mid-size tablets, and large tablets are widely available for sale today, and all demand different approaches to software design because of ergonomics, form factors, and general patterns of use (such as joint ownership of larger tablets). These differences are discussed in Chapter 3 "Android Fragmentation," and are considered throughout this book.

However, phones and tablets are not the only platforms you need to worry about. Futuristic installations of Android OS in everything from ski goggles to refrigerators and cars are only a matter of time. All these installations require considerable changes to the user interface to fit the specific needs of the device. That means that the old model of wireframing no longer reflects the rich and variable reality on the ground. In order to understand design constraints for these objects, you need to modify your design approach to include the physical form factor of the device, as well as transient elements such as animations and transitions.

Your User Testing Must Allow People to Explore the Natural Range of Motion, Voice, and Multitouch

When it comes to mobile design and testing, forget everything you know about interacting with a computer. The uniform mode of interacting with a computer via only mouse and keyboard does not apply to mobile devices. Much of what the mobile age is all about is taking advantage of the body's natural motions: scratching to dig deeper, shaking to say no, and bringing the phone to your ear to speak. From voice recognition digital assistants to pedometers that use the body's swaying motion along with GPS on-board sensors to determine the speed and quality of daily physical activity, today's mobile devices are using an unprecedented array of motion, voice, and multitouch gestures to obtain increasingly complex inputs from customers. In order to design effective interfaces, your prototype and the customer experience testing techniques need to account for a full range of these new modes of interaction with the devices.

Touch Interfaces Embody Simplicity and Sophistication

Browser and OS-based software that runs on large screens could afford to have ill-conceived advertising modules while still succeeding in converting customers because of the high tolerance of complexity, fairly large screens, and relatively high degree of focus that customers had on your software just by virtue of having to sit down at a chair to use the computer.

The mobile age is all about, well, mobility. That means your customers' attention is even more scattered than ever before, more, even, than anyone thought possible just five years ago. This means that interfaces need to be simple. That's not to say "simplistic"—as Edward Tufte famously said, there is an ocean of difference between simplicity and simple-mindedness. Instead, the *software needs to take on the burden of complexity that was heretofore acceptable to pass on to the consumer.*

Don't misunderstand: People actually want to do *more* with smartphones and tablets than they could do with the web. You simply can't afford the *perception* of complexity anymore. So in touch interfaces, you have a unique customer interface that is *not complicated, but very sophisticated*. This means that a device's touch interfaces are, in many respects, easier to prototype than desktop web interfaces—especially when it comes to low-fidelity methods like paper prototyping—as long as the moderator makes an effort to probe the less tangible aspects of the interface.

Delight Is Mandatory

Delight, fun, and games existed under the old PC and Mac system. However, the majority of "fun" was devoted to specific activities, such as computer games. The new mobile platform has grown up on games. The games are in its blood and DNA. Thus, no matter how boring and trivial the task, designers need to make sure that the software is as delightful as possible, even if "delight" simply means the software helped the customer complete the task as quickly as possible.

Increased gamification is a natural outcome of the new platform, and, as John Ferrara says in his book *Playful Design* (Rosenfeld Media, 2012), the experience of play must be delightful as a stand-alone activity, not feel tacked on as an afterthought onto some other agenda. This means that the best mobile experiences must behave and feel more like games. The small size of the screen makes it inevitable, for example, that fun elements (such as transitions) play a big role in the experience. In contrast, the old browser model had minimal transitions. This means that while you are prototyping the design of your app, you must allow time to explore the elements of transitions, delight, and gamification.

Tell a Complete Story—Design for Cross-Channel Experiences

Using a PC or Mac was almost always work. With the exception of a few hundred super-geeks who took their computers to the restroom with them and refused to shower so that they could spend as little time as possible offline, most people would have dedicated "online" or "computer" time to accomplish digital tasks. In contrast, many "normal" people have their mobile devices with them at all times. More and more, there are increasing numbers of people that sleep, eat, and even go to the restroom (shocking!) with their digital devices. Because of the incredible array of on-board sensors (microphone, GPS, light sensor, camera, near field communica-tions [NFC], touch, motion, and so on) the mobile experience creates a completely unprecedented connection between offline (also known as "real-world") and digital realities. It's as though we have acquired a new organ that enables us to connect to the unseen digital worlds of Facebook, read QR codes and NFC chips, and access interconnected digital information such as maps and reviews in the moment we need the information. This new "mobile organ," which is always with us, is a com-pletely different way to easily and quickly access and manipulate information.

Today, it is safe to assume that the "mobile organ" will be used to augment every traditionally offline experience, such as a visit to an amusement park, a shopping trip, and even a hike in the woods. As a designer, you need to pay close attention to the *spaces between* interactions, where a mobile device is used to interact with other channels. For example, someone might start a task on a mobile device, continue on the desktop and social networks, and finish it in the physical store. These quick tasks that appear to be done "on the side," at the spur of the moment, or while wait-ing for some other event to occur might just be your primary mobile use cases.

Now that you know some of the challenges you are facing, the following section includes a mobile design case study that will help you understand how you can put all this information together into a User-Centered Design (UCD) process that works for mobile.

Mobile Design Case Study

I used a lightweight Agile mobile design process to help deliver a radical innova-tion: an authentically mobile "60-second listing" flow for the ThirstyPocket iPhone app, which will also be shortly coming to the Android Marketplace. This project is an illustrative case study of how to apply UCD to mobile design. This case study is provided merely as an illustration of some of the concepts discussed in the pre-ceding sections, such as lightweight prototyping. You may have to adjust your design approach and process based on your particular situation. The key is to stay flexible while remaining customer-focused.

Step 1: Scope, Concept, and Planning

Before proposing any design solutions, start by having a kick-off meeting to understanding the who, where, how, and how much questions, otherwise known as context, persona, vision, and budget. Depending on the project, this could be as simple as writing these in a one-sentence format. It is critical that the entire team agrees to these four points, and any questions are called out, so they can be answered through research.

Context: Where Will the Product Be Used?

As noted above, context is the key to creating a great mobile design. You must know where your potential customers are, what equipment they use, what else they are doing at the time of interacting with your app, and what emotional state and concerns they have about the process. All of these factors will help drive the design and ultimately determine your choice of a final design approach and the list of features and system behaviors your product should have. During the planning and scoping phase, you need to write down the team's current thinking to have a starting point for customer research. For the ThirstyPocket app, the context was "Garage sale in an urban area."

Persona: Who Is the Target Customer?

You may start out knowing this, or the idea of a target customer may evolve from the internal discussion or field research. Regardless of how you arrive at the insight, it is essential to have an agreement of who the solution is supposed to target, even if your assumption is not correct. If the team disagrees on the target persona, go ahead and document this—disagreements point to areas where more research is needed. Although recommended, you don't need to spend a lot of time developing sophisticated, detailed personas for your mobile project. Sometimes a one-sentence "persona sketch," such as a simple "young college student, not a lot of money or time" that we agreed upon for ThirstyPocket, could be enough to jump into testing. The most important function of the persona is the sense of team cohesion and empathy toward the struggles and challenges faced by the target customer. If you are feeling strange about coming up with a persona sketch without a lot of information, keep in mind that having fictional personas are better than none. At least you will have documented your team's assumptions, so you can quickly discover if they are not correct and update as necessary when you jump into field research.

Field Research and Contextual Interviews

After the context and persona sketch is complete, jump into testing! For ThirstyPocket, we visited plenty of garage sales and interviewed people who fit the target profile about selling their stuff to get a better understanding of the challenges of the existing selling systems and workflows and people's frustrations with the process. As you do your research, be sure to go on field studies together with other team members and discuss findings and brainstorm ideas immediately after the research sessions. This is especially great to do over a coffee, lunch, or dinner following the research session. No great documentation is needed—simple paper sketches and idea drawings shared in context with the entire team are often the best ways to come up with great ideas, improvements, and product vision. Don't forget to test your assumptions and correct the persona sketch and context understanding as necessary.

Vision: How Will the Product Be Used?

How do you envision the product being used? Is it a long engagement or a quick information snack on the run? How often does the customer engage? What triggers the engagement? What is the service window when the customer interacts with your app? Does the interaction span multiple touch points, and must customers come back at some later time? Does your app require preparation or training? What happens front stage, between customers? What happens back stage on the software or service side of the equation? Ultimately, your job is to understand and sharpen your complete vision as part of your design process. For the Thirsty-Pocket app, we came up with a cool slogan: "List your item in 60 seconds." Your vision for how the product will be used should ideally be rooted in direct team observation of the target customers in target context. From this point, it is very easy to jump into the next section, "Step 2, Design Workshop."

Budget: How Much Time and Money Do You Plan to Spend on Design and Development?

UX design is only a small piece of the product development puzzle. Take the time to understand how the design fits into your overall development plan and work within your timelines to make the most of your team's technical capabilities. A typical design process takes anywhere from 3 to 6 months. For the ThirstyPocket app, the budget was set for 3 months.

After you've established your context, persona, vision, and budget, you can move on to the design workshop described in the next section.

Step 2: Design Workshop

At the start of the workshop, and before you propose any design solutions, you must first focus on nailing down four essential pieces of information: personas, context, scenarios, and vision. You focus on driving the shared team-wide understanding and buy-in by developing use-case scenarios and a beefed-up vision statement, which help you fill in any missing pieces in the framework you developed in the first step. In the case of the ThirstyPocket app, use case scenarios and vision were updated as follows:

- **Scenarios:** Selling a car in the neighborhood, around the 5-mile range; selling tickets at the last minute while at a concert using texting/phone for communication.

- **Enhanced Vision:** Local, social, e-commerce. Seeing the product in person and paying in cash, like a garage sale. No shipping. Only a single picture. Natural, easy, and simple selling process, with no registration/login.

After you nail down the four essentials, and often during the process, you generate ideas by working collaboratively as a team.

It is a good idea to always approach the design process with multiple directions in mind, injecting a rigorous discipline into the brainstorming process. Rather than getting excited about the particular approach, document it quickly and then set it completely aside, asking the following questions:

- "Is there another way to design this?"

- "What if you start with X instead of Y?"

- "Can you make it more like X?"

- "Can you make the new workflow fit the existing customer behaviors?"

- "Can you make this workflow feel more like a game?"

Storyboarding is an excellent technique to document various design ideas in context and as they happen through time. The key is to keep the actual interface design portions to the minimum and instead describe how the app will be used in the mobile context. For ThirstyPocket, the approach shown in Figure 4.1 fit well with both scenarios, and it fulfilled the larger product vision.

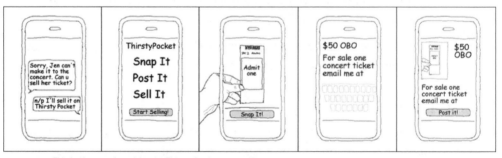

FIGURE 4.1: This is the storyboard for the ThirstyPocket app selling process.

For example, the storyboard documents the following scenario: A young person named Gene is heading to a cool concert in a car with a bunch of friends. Suddenly he gets a text—his friend Jen can't make it! Could Gene sell her ticket? "No problem," replies Gene, "I can sell it on ThirstyPocket." This opening storyboard is important because it shows the context of interaction. It should convey something about the scenario as well as setting up a strong sense of place where the scenario is happening.

Gene next fires up the ThirstyPocket app ("Snap it, Post it, Sell it!"), and proceeds to tap Start Selling! This action opens the built-in camera, and Gene taps the Snap

It! button to take the picture of the ticket he is looking to sell. Gene fills out a quick description and taps the Post It! button on some kind of a preview screen. And there you have it—a simple, quick, locally focused selling flow without a great deal of interface details.

During the workshop you concentrate on speed, quickly exploring various design scenario storyboards on a whiteboard or with small rectangular sticky notes. You must strive first and foremost for mutual understanding and strong vision, not exhaustive documentation. A full description of storyboarding techniques is outside the scope of this book. Check out the incomparable *Making Comics* (Harper, 2006) by Scott McCloud, and visit the companion site for this book, http://AndroidDesignBook .com, for more examples of mobile storyboards and use cases.

During the workshop, encourage everyone on your team to draw and participate. High-value production storyboards are not necessary. If a whiteboard full of chicken scratches surrounding stick people is enough for the entire team to "get" the particular mobile scenario, that's all you need. After you have the key use-case scenarios story-boarded, it's time for "Step 3: Wireframe and RITE Study with Sticky Notes."

Step 3: Wireframe and RITE Study with Sticky Notes

As discussed earlier in the chapter, due to unusual design constraints of mobile, the commonly used UCD process of creating computer-generated wireframes and then building a high-fidelity prototype does not always work for mobile design.

Instead of spending a lot of time and effort creating high-fidelity wireframes, set up a cheap and efficient Rapid Iterative Testing and Evaluation (RITE) study as the core of your design process. Do RITE studies as early as possible in the design process and you will reap the benefits of creating more delightful, usable, and successful mobile products in less time than you ever thought possible.

The RITE study (you can call it "RITE test" if you like, though I prefer the term "study" to emphasize the design changes) I typically recommend is conducted using 9 to 12 participants in three to four rounds (3 participants per round). *The critical component of a RITE study is to allow time between rounds for updating the prototype to fix the issues discovered during the previous round's testing.* Basically a RITE study is a series of design/test pairings where the prototype is rapidly changed as needed based on feedback from the customers, engineers, and management.

RITE studies have been a part of the UCD toolbox for many years. One simple modification I came up with to enable RITE to work well as a core of the mobile design process is to employ the prototype made from sticky notes.

Mobile sticky-notes prototypes offer many advantages. To begin with, a pack of large sticky notes (I prefer to use 3 × 5 inch size) has the dimensions that resemble those of a typical mobile phone. That means that there is no need to create any external cases, boxes, or anything else to resemble the mobile device; the pack of sticky notes is a complete solution.

Sticky-note prototypes are cheap, are easy to create, and they are fairly robust. They can be dropped from any height without disintegrating or even so much as falling apart into individual pages. You won't have any issues handing a pack of post it notes to a complete stranger on the street or in the coffee shop and asking him some questions about your app (most people would not have the same light-hearted feelings about handing their precious late-model mobile phone to a complete stranger). If the participant happens to accidentally drop the sticky-note "phone," it will not be damaged, and if they happen to run off with it, you will only be out about a dollar!

As shown in Figure 4.2, a pack of sticky notes closely resembles the form-factor of the mobile device, so this simple yet sophisticated prototype enables you to test the natural ergonomics, multitouch, and accelerometer motions—something traditional wireframes simply cannot accomplish.

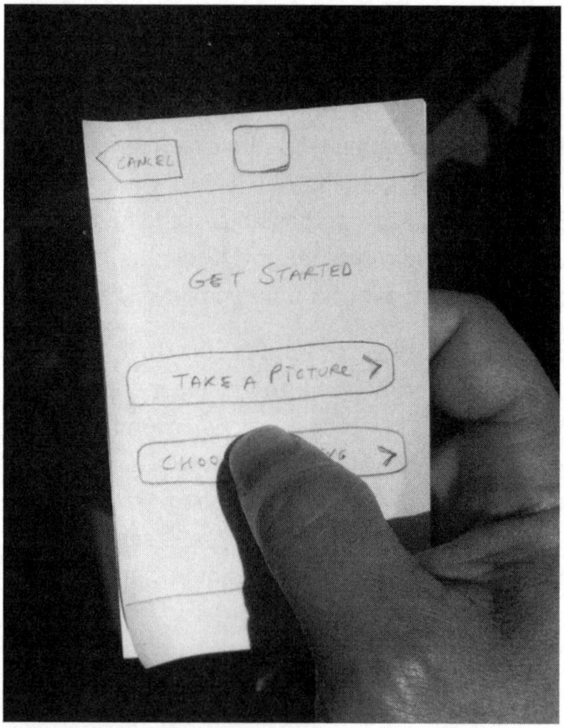

FIGURE 4.2: Using a pack of sticky notes to simulate a phone is an effective, lightweight prototyping technique.

Sticky-note prototypes are easy to fix: If you discover issues with the design, you can fix the interface right then and there, using an eraser and a pencil, or a new sticky note re-drawn with a fine-point permanent marker. Likewise, if you want to test an alternative flow, you can draw a new screen design in a few minutes and compare its performance with that of an existing idea almost immediately, when you hand it to the next evaluator. This ease of use helps your design evolve quickly, especially if your entire core team is present at these test sessions.

Figure 4.3 shows how I used the sticky-note method to wireframe the original "Sell your item in 60 seconds" flow I tested for ThirstyPocket.

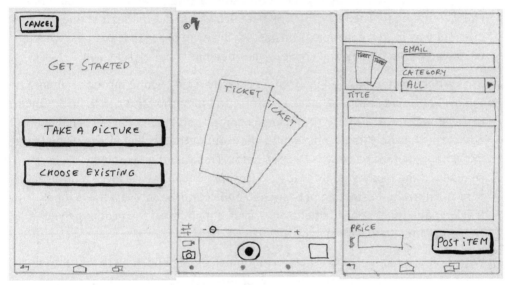

FIGURE 4.3: This is the mobile design prototype for early testing that I created using sticky notes.

Because I have personally experienced excellent results using my sticky-note prototype approach, and I firmly believe that it will help you reach your design goals, I have created 3 × 5 inch sticky note wireframes for most of the design patterns in this book. As you explore this prototyping technique, here are a few things to keep in mind:

- When using the sticky-note pack to approximate the mobile device, you do not need to draw an additional box on top of the sticky note to represent the screen. To save time, as well as to make the drawing more understandable, go ahead and assume that the entire sticky-note surface is the screen of the mobile phone. For Android phones you should add the device's hardware buttons (Return, Home, etc.) where appropriate.

- All drawings in this book are black and white, executed using Pigma Micron pens with Archival Ink. This is done for legibility. In real-life prototyping I use a simple mechanical #2 pencil so that I can immediately erase or update any elements on the "screen." You can use any pen or pencil you like, black or color.

- I use a ruler liberally while drawing my wireframes. This is because drawing a straight line, especially while making rapid changes in the field, can be a bit of a challenge. Whether or not you decide to use a ruler, it's best to be consistent in your technique throughout the prototype, so as not to distract your study participants. There are various templates and other drawing aids out there to help you. I am not a fan of anything other than a small transparent triangular ruler. Remember: The key is not the best drawings, but the fastest and most efficient way to test concepts. The easiest drawing process is the best. Use whatever works for you and conveys your designs most effectively.

- Sticky-note prototypes can be branched easily. This is done through handing the evaluator only the first "screen" stuck on top of the sticky-note pack. Once the evaluator taps a button or performs some function on that screen, I can select the appropriate *next* screen to show out of the pack of alternatives I hold in my hand, out of the view of the evaluator. This way the testing can be very realistic: If the next evaluator taps a different control, he receives a different screen, allowing for testing of branched and round-about workflows, back-tracking and other real-life behaviors. This capability of the prototype yields rich and robust behavioral data.

- With a bit of practice, sticky-note prototypes can also be used to test transitions. If transitions are important to your interaction, try "slipping in" the next "screen" while saying something such as, "Let's say the next page comes up from the bottom like so. How does that feel for you?" If the participant responds with something noncommittal such as, "Fine," ask her if she would prefer a different transition instead, or if this movement has some sort of meaning to her. You might need to do a complex transition more than once to convey it properly. You may also wish to draw individual transition states on separate sticky notes (see "Storyboarding iPad Transitions," *Boxes and Arrows*, January 5, 2011, http://www.designcaffeine.com/articles/storyboarding-ipad-mobile-transitions/).

- Keyboards are easily mocked up by having another smaller sticky note that can be overlaid on top of the sticky-note pack "device" the study participant is holding. This way any screen can be converted into the "keyboard entry screen" dynamically at any time during the test. This technique gives additional flexibility to the prototype while removing the chore of drawing a complex key-board design element over and over again.

- Remember to revisit your storyboard. Your wireframe's workflow should follow your product vision. For example, compare Figure 4.1 with Figure 4.3; you can see that the wireframes are a natural extension of the original vision with a few more details and interface elements added. For example, shortly after I started the testing, it became apparent that people sometimes wanted to take a bunch of pictures first and sell items later. Thus, I added another screen that gave the customer a choice: Take a Picture or Choose Existing (picture) (refer to Figure 4.3). These kinds of changes are common. They are the reason you want to test your idea on paper in the first place! If the change is drastic enough, you must update your vision storyboard accordingly after you learn new insights during your RITE study. In this case, the additional page did not change the basic storyboard scenario—it merely added fidelity to it—so there was no need to change the original vision storyboard (refer to Figure 4.1).

Sticky-note prototypes enable you to quickly and inexpensively explore multiple design approaches while dispensing with elaborate camera equipment and other gadgets. Sticky-note prototypes also enable you and your entire team to escape the confines of the office and boldly go to where the mobile interaction is actually taking place: coffee shops, busy street corners, taxis, and subways. *There is tremendous value in putting your paper prototypes in the hands of potential customers* in situ, *on location where the interaction for which you are trying to design is actually taking place.*

At the same time, study participants can be comfortable brainstorming valuable ideas because the prototypes do not have that "finished" look. This leads to powerful collaborative "jam" design sessions with the customers that take place directly in the context where they would be using your creation. These sessions yield invaluable insights you directly incorporate into your designs, much faster than you ever thought possible. *That is why I included hand-drawn sticky-note wireframes with nearly every design pattern in the book.* Hopefully, these will give you an inspiration to build your own sticky-note prototypes and test them in context, directly with your target customers. If you need help, please visit the companion site for this book (http://AndroidDesignBook.com) where I host videos of example hands-on RITE studies using sticky-note prototypes, mobile use case storyboards and many other resources designed to help you get the most out of this powerful technique.

Step 4: Visual Design

A RITE study is not the final deliverable; it's merely a key step in the design process. The best way to think about your design process is that *the state of your prototype and deliverables should reflect the overall state of completion of your product.* That's why early in the process, you will be able to move quickly by employing a lightweight design process to focus on designing the rough flows and screen layouts that work for the customer. After the rough flows have been worked out and tested, it's time to move on to the final step: visual design.

It is customary to employ visual designers and content managers to give the final project that showroom luster. Although the visual design is not the focus of this book, Chapters 1 and 2 include a few essential pointers on the subject.

It is worth noting that visual design can both enhance and detract from the interaction design intent. Sometimes visual design can make a big difference. Thus it is a good idea to test the app a few more times to ensure that, despite the evolution of the design, the final version remains true to the original vision storyboard's simplicity and elegance. Styling can be key to creating (or destroying) emotional connections, so it can be worth testing for that as well.

To run the final tests, simply hand the testing device (careful now!) to the person whose opinion you are interested in, perhaps while waiting in line at a coffee shop. Say, "Let me know what you think about the app, and I'll buy your morning coffee." This final testing of five to eight people should take no longer than an hour or so. This is an excellent test: Your final flow should ideally take no longer than it does for a person to reach the coffee counter during the morning rush, and the interface should be engaging and self-explanatory enough for the person to manipulate while in the decaffeinated state of mind.

PART II

Android Design Patterns and Antipatterns

CHAPTER 5

Welcome Experience

The first thing your customers see when they download and open your app is the welcome mat you roll out for them. Unfortunately, this welcome mat commonly contains unfriendly impediments to progress and engagement: End User License Agreements (EULAs), disclaimers, and sign-up forms. Like the overzealous zombie cross-breeds between lawyers and customs agents, these antipatterns require multiple forms to be filled out in triplicate, while keeping the customers from enjoying the app they have so laboriously invested time and flash memory space to download. This chapter exposes the culprits and suggests a few friendlier welcome strategies that keep the lawyer-custom agent crossbreed zombies at bay.

⊖ 5.1 Antipattern: End User License Agreements (EULAs)

When customers open a mobile website, they can often engage immediately. Ironically, the same information accessed through apps frequently requires agreeing to various EULAs, often accompanied by ingenious strategies that force customers to slow down. EULAs are antipatterns.

When and Where It Shows Up

EULAs are typically shown to the customer when the application is first launched and before the person can use the app. Unfortunately, when they do show up, EULAs are also frequently accompanied by various interface devices designed to slow people down. Some EULAs require people to scroll or paginate to the end of a 20-page document of incomprehensible lawyer-speak before they allow access. Others purposefully slow people down with confirmation screens that require extra taps. Truly, things have evolved nicely since the days of medieval tortures!

Example

Financial giant Chase provides a good example of a EULA. As shown in Figure 5.1, when customers first download the Chase app, they are faced with having to accept a EULA even before they can log in.

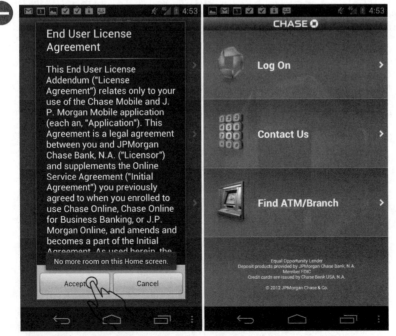

FIGURE 5.1: The Chase app includes the EULA antipattern.

What makes this example interesting is that the same information is accessible on the mobile phone without needing to accept the EULA first: through the mobile web browser, as shown in Figure 5.2.

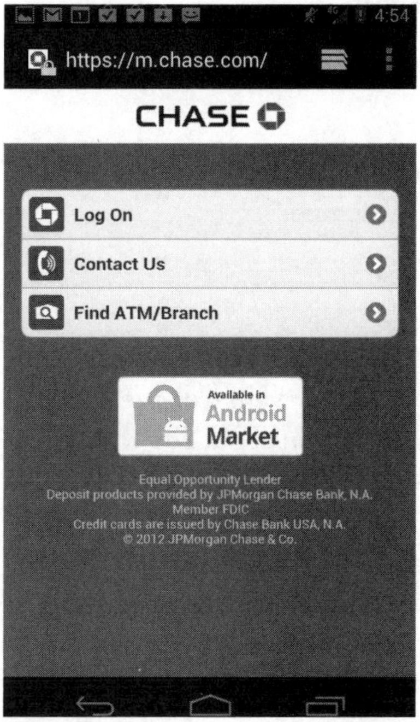

FIGURE 5.2: There is no EULA on the Chase mobile website.

Why Avoid It

The remarkable thing is *not* that the EULA is required. Lawyers want to eat, too, so the EULAs are an important component of today's litigious society. Dealing with a first-world bank in the "New Normal" pretty much guarantees that you'll be faced with signing some sort of a legal agreement *at some point* in the relationship. The issue is not the EULA itself—it is the *thoughtlessness of the timing* of the EULA's appearance.

The app has no idea if you have turned mobile access on or have your password set up properly. (Most people have at least a few issues with this.) Therefore, the app has no idea if the bank can serve you on this device. However, already, the bank managed to warn you that doing business on the mobile device is dangerous and foolhardy and, should you choose to be reckless enough to continue, the bank thereby has no reasonable choice but to relinquish any and all responsibility for the future of your money. This is hardly an excellent way to start a mature brand relationship.

What should happen instead? Well, the mobile website provides a clue. First, it shows what the user can do without logging in, such as finding a local branch or an ATM. Next, the mobile site enables the customer to log in. Then the system determines the state of the EULA that's on file. If (to paraphrase Cream's lyrics in "The Tales of Brave Ulysses") the customers' "naked ears were tortured by the EULA's sweetly singing" at some point in the past, great—no need to repeat the sheer awesomeness of the experience. If not, well, it's lawyer time. Consequently, if customers do not have Bill Pay turned on, for example, they don't need to sign a Bill Pay EULA at all, now do they?

The point is that the first page customers get when they first launch your app is your welcome mat. Make sure yours actually says "Welcome."

Additional Considerations

Has anyone bothered asking, "How many relationships (that end well) begin with a EULA anyway?" How would your use of the Internet feel if every website you navigate to first asks you to agree to a EULA, even before you could see what the site is about? That just does not happen. You navigate to a website and see engaging welcome content immediately. (Otherwise, you'd be out of there before you could spell E-U-L-A.) When you use a site to purchase something, you get a simple Agree and Proceed button with a nearby link to a EULA agreement (not that anyone ever reads those things anyway) and merely proceed on your way.

If you can surf the web happily, taking for granted the magnificence of the smorgasbord of information on the mobile and desktop, without ever giving a second thought to the EULAs, why do you need to tolerate a welcome mat of thoughtless invasive agreements on the world's fastest growing platform?

Related Patterns

5.2 Antipattern: Contact Us Impediments

⊖ 5.2 Antipattern: Contact Us Impediments

Whenever things go wrong, your customers need to contact support. Including impediments such as untappable phone numbers and long contact forms without presets is a User Experience (UX) antipattern.

When and Where It Shows Up

This antipattern is unfortunately quite common because it takes time and effort to mobilize the customer support contact mechanisms. These often get left behind when desktop web applications are converted into mobile apps.

Example

In the US Bank app shown in Figure 5.3, things went wrong during the initial registration. At this point in the flow, customers are completely stuck. There is literally no self-service possible—customers *must* call customer support. So the app notifies customers of the error by a pop-up alert, asking them to call the number shown. Unfortunately, the phone number is not tappable. Nor is it possible to copy the phone number. In other words, customers are forced to either remember the phone number they need to dial or else they need to get a piece of paper and write down the number. Don't forget that the device your customers hold is also a phone. Forcing customers to write down the number they need to call *while they are on the mobile phone already* is an egregious failure of service—you might as well ask them to chisel it on a little stone pyramid or write it in plant pigment on the walls of their cave.

FIGURE 5.3: When things go wrong during registration in the US Bank app, the customer experiences the Contact Us Impediments antipattern.

However, you can argue that a simple act of having to write down a phone number simply pales in comparison with sin #2: long Contact Us forms that are devoid of any presets. The example in Figure 5.4 is courtesy of Kodak.

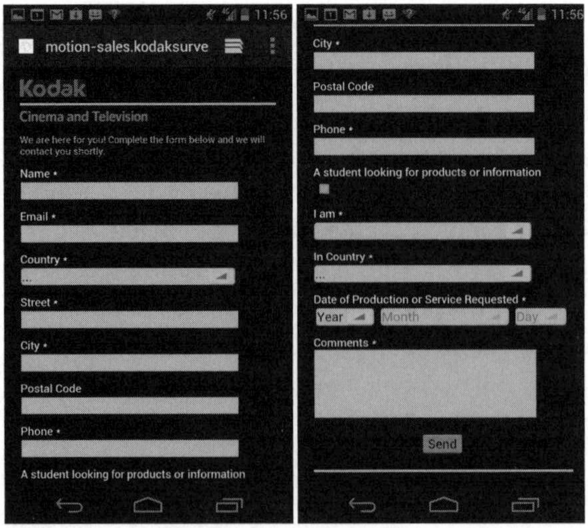

FIGURE 5.4: The Contact Us Impediments antipattern is expressed as a long form in the Kodak app.

Yes, sure, I'd like more information about your product. But no thank you; I will not fill out a helpful form that has more fields than an average mobile customer fills out in a week.

Why Avoid It

At the point the customer needs to seek technical assistance, she is already unrecoverably stuck, which means that she is already frustrated with your app, your company, and your brand. Throwing up impediments and barriers to contacting the people she needs to get to simply does not make sense—it's an antipattern.

Additional Considerations

Frequently, I hear from my consulting clients the argument that allowing their customers to contact tech support by phone is just too costly. On the other hand, these companies need to allow access to the phone number or e-mail address to enable their customers to recover in cases where they cannot do so simply using the online help. So the companies compromise by giving a number but making it difficult to dial with the single-tap action the phone makes so easy—in other words, the companies create Contact Us Impediments. This is a false economy. Do not forget that the smartphone is also a phone and an e-mail client. If the customer is forced to do something unnatural with these simple calls to action when she is in trouble and needs help, your brand will be damaged in significant ways, and the customer will be that much harder to deal with when she finally calls your 800 number after she manages to write it down. If you don't want your customers calling you, simply do not provide a number.

Related Patterns

9.2 Antipattern: Lack of Interface Efficiency

⊖ 5.3 Antipattern: Sign Up/Sign In

I hope by now you are getting my drift: Anything that slows down customers or gets in their way after they download the app is a bad thing. That includes sign-up/sign-in forms that show up even before potential customers can figure out if the app is actually worth using.

When and Where It Shows Up

This antipattern seems to be going away more and more as companies are beginning to figure out the following simple equation:

Long sign-up form before you can use the app = Delete app

However, a fair number of apps still force customers to sign up, sign in, or perform some other useless action before they can use the app.

Example

The application SitOrSquat is a brilliant little piece of social engineering software that enables people to find bathrooms on the go, when they gotta go. Obviously, the basic use case implies, shall we say, a certain sense of urgency. This urgency is all but unfelt by the company that acquired the app, Procter and Gamble (P&G), as it would appear for the express purposes of marketing the Charmin brand of toilet paper. (It's truly a match made in heaven—but I digress.)

Not content with the business of simply "Squeezing the Charmin" (that is, simple advertising), P&G executives decided for some unfathomable reason to force people to sign up for the app in multiple ways. First, as you can see in Figure 5.5, the app forces the customer (who is urgently looking for a place to relieve himself, let's not forget) to use the awkward picker control to select his birthday to allegedly find out if he has been "potty trained." This requirement would be torture on a normal day, but—I think you'll agree—it's excruciating when you really gotta go.

But the fun does not stop there—if (and only if) the customer manages to use the picker to select the month and year of his birth *correctly* (and how, exactly, does the app know the customer has selected correctly?) he then sees the EULA, which, as discussed earlier in the chapter, is an antipattern all to itself. The EULA is long, complex, and written in such tiny font that reading it while waiting to go to the bathroom should be considered an Olympic sport, to be performed only once every 4 years. Assuming the customer gets through the EULA, P&G presents yet another sign-up screen, offering the user the option to sign in with Facebook. I guess no one told the P&G execs that the Twitter message "pooping" is actually a prank. They must have legitimately thought that they could transfer some sort of social engineering information about the person's bathroom habits to "achieve and maintain synergistic Facebook connectivity." I would have to struggle hard to find monumental absurdities from social networking experiments that are equal to this. I can't imagine that anyone thinks, "Finally! Sharing my bathroom habits on Facebook has never been easier!"

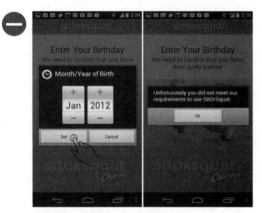

FIGURE 5.5: This registration failure is a Sign Up/Sign In antipattern in the SitOrSquat app.

Assuming that the user is a legitimate customer looking to use the bathroom for its intended purpose, and not a coprophiliac Facebook exhibitionist, the assumption is that he will naturally dismiss the Facebook sign-in screen and come to the next jewel: the tutorial. Note that the entire app *outside of registration* consists of basically four screens if you count the functionality to add bathrooms. However, if you include all the sign-up antipattern screens (including my initial failure to prove that my potty training certificate is up to date, as referred to in Figure 5.5), it takes *seven screens* of the preliminary garbage before the content you are looking for finally shows up (see Figure 5.6). If you count the number of taps necessary to enter my birthday, there are almost *50 taps*!

One of my favorite UX people, Tamara Adlin (who is coauthor of *The Persona Lifecycle: Keeping People in Mind During Product Design* with John Pruitt, 2006, Morgan Kaufmann) wrote brilliantly: "For Heaven's Sakes, Let Them Pee." I believe that never before has this line been so appropriate. In the absurd pursuit of social media "exposure" coupled with endless sign-up screens, and heavy-handed "lawyering up," P&G executives completely lost sight of the primary use case: letting their customer SitOrSquat.

FIGURE 5.6: Behold the glory of the Complete Sign Up/Sign In antipattern in the SitOrSquat app.

Why Avoid It

Long sign-up screens detract from the key mobile use case: quick, simple infor-mation access on the go. Overly invasive sign-up/sign-in screens presented up front and without due cause can cause your customers to delete the app.

Additional Considerations

When deciding whether to force the customer to perform an action, consider this: If this were a web app, would you force the customer to do this? If you have an Internet connection, you can save everything the customer does and connect it back to his device using a simple session token and a guest account. And even if you don't (for example, while riding in a subway, using airplane mode, and so on), today's smartphones have plenty of on-board storage you can use for later sync-ing with your servers when the mobile network eventually becomes available.

This means *there is simply no reason to force anyone to register for anything,* other than if they want to share the data from their phone with other devices. As a gen-eral rule, rather than forcing registration upon download or at the first opportunity, it is much better to allow the customer to save a piece of information locally on the phone without requiring that he log in. Wait until the customer asks for something that requires registration, such as sharing the information with another device or accessing information already saved in his account; at that point completing the registration makes perfect sense.

For example, imagine how absurd the Amazon.com shopping experience would be if the app asked you for your home address, billing address, and credit card *upfront*—before allowing you to see a single item for sale! Yet entering the home address (where would you like to have the items shipped?) and credit card (how would you like to pay for this?) makes perfect sense during the checkout, *after* the customer selects a few items and indicates she would like to complete the purchase.

Finally, as Luke Wroblewski quipped in his book *Web Form Design* (Rosenfeld Media, 2008), "Forms suck." Only ask for what you strictly need to proceed to the next step and omit extraneous information. (Read more detail on this topic in Chapters 10 through 12.)

Related Patterns

None

5.4 Pattern: Welcome Animation

Welcome animation is a prominent feature on iPhone; however, it's all but missing on the Android platform. As Reel 2 Real sang in a (very) animated film *Madagascar*, it's time to "Move it, move it."

How It Works

When the customer first opens the app, a small animation clip plays to welcome the customer and show off the brand. Often this is done tongue-in-cheek, with a bit of light humor.

Example

Android boot animations are often lavish and spectacular affairs, such as the one for the Galaxy Nexus Ice Cream Sandwich OS, as shown in Figure 5.7.

FIGURE 5.7: This lavish boot animation variant of the Welcome Animation is on the Galaxy Nexus.

However, animations for apps seem to be mostly absent on the Android, so I had to look to iPhone to provide a suitable example. One of my favorite implementations of this pattern is the Priceline app on the iPhone. (See Figure 5.8; the app does not exist on Android at the time of this writing.) You can hardly do better than having Cap'n James Tiberius Kirk punching a hole through your phone; it certainly gets your attention!

FIGURE 5.8: The Priceline iPhone app includes a creative implementation of the Welcome Animation.

When and Where to Use It

Apps that have longer startup times can use a Welcome Animation to help occupy people while they wait for the app to load. However, as startup times become shorter, this pattern is used more for effect and branding purposes, and it runs *only once* when the app is installed.

Why Use It

Smartphones are used to have fun, play games, and accomplish serious tasks. (Once in a while, people even use them for making a phone call.) A Welcome Animation is a way to set the atmosphere for a fun experience, not only for games, but also for serious apps. A Welcome Animation is also an excellent way to pass the time when you have a slow transition.

Other Uses

One of the best ways to use the Welcome Animation is to tell a story. The best apps create a story that is integrated as a tutorial at the same time (see the "Tutorial" section for examples).

Pet Shop Application

Welcome to the Pet Shop application! This important section is a homage to the *Java Pet Store*, the original reference application for the Java 2 Enterprise Edition (J2EE) platform created by Connie Weiss and Greg Murray (http://en.wikipedia.org/wiki/Java_BluePrints). Thousands of people (including the author) learned how to program in Java by looking at the code patterns in the Java Pet Store. Throughout this book, I follow a similar format, presenting various Android patterns using hand-sketched sticky-note wireframes.

Ironically, I'm starting with the section that does not have a wireframe. Why not? Two reasons: First, it's easy to imagine a snazzy Welcome Animation of happy cartoony dogs or cats running or jumping around the screen, so you don't really need to draw a wireframe for it. Second, actually drawing a Welcome Animation in the production-ready format is just a bit beyond the drawing capabilities of most people (including yours truly).

Rather than presenting some spectacular welcome drawing and thereby discouraging anyone reading this book from trying to draw, I chose to omit the drawing entirely. In this book I stick to simple wireframes that *anyone should be able to draw*. Indeed, drawing your own wireframes and creating practical sticky-note prototypes for customer testing based on the patterns I present here *is the entire point of this book*.

That said, if you would like to give it a shot to see how a cartoony cat and dog Welcome Animation might look, feel free to draw one using the sticky-note methodology I outline in Chapter 4. (If you do, be sure to send it to my team at http://androiddesignbook.com. I would love to feature it alongside our tutorials.)

Tablet Apps

This pattern should work the same way for tablets as it does for the smaller mobile phones, just be mindful of screen resolution. You want to make sure the animation looks great and fits the screen; otherwise, it defeats the entire purpose and you are better off skipping it.

! Caution

In addition to the warning about keeping screen resolution in mind when designing for tablets, two things you should keep in mind are

1. Don't make the animation too long—3 to 5 seconds is plenty. Anything longer than 10 seconds starts to feel too long.

2. Don't run the animation more than once. It should run only when the app is launched, if needed. Don't run it when switching between apps during multitasking.

Related Patterns

5.5 Pattern: Tutorial

5.5 Pattern: Tutorial

In my workshops, I am often asked about designing tutorials. Most apps don't need one, but occasionally, apps that have nonobvious features need a little help explaining how to get the most out of the functionality.

How It Works

During the welcome experience, the customer gets a short lesson in how to use the app.

Example

There are many spectacular examples of tutorials. One format uses a separate page that explains how the app functions. The SitOrSquat app example includes a tutorial of this type (refer to the "5.3. Antipattern: Sign Up/Sign In" section earlier in this chapter). Although this style is the easiest and cheapest to implement, "extra page" tutorials are also the most annoying and frequently unnecessary because they directly interrupt the natural flow of engagement with the app.

In contrast, the best tutorials are integrated directly into the use of the app. My favorite examples come from games such as the N.O.V.A series by Gameloft, where tutorials are so much a part of the story that they can be considered a work of art. Note the prominent "Skip" button in Figure 5.9.

When the person starts playing the game for the first time, Yelena, the virtual assistant AI comes on and asks you to "check your guidance system" by using it to turn your head and walk around. It then performs similar "guided checks" of the weapons system while giving you helpful notes in snide tones perfectly in keeping with the rich storyline of the game.

Fortunately, you don't need to spend a million dollars or know anything about storytelling to create an integrated tutorial. All you need is a simple overlay, also called a Watermark, which is covered in more detail in Chapter 13, "Navigation." A good example of using overlays for tutorial purposes is the Flipboard app (see

Figure 5.10), which features a simple overlay pointing to the ribbon-like window shade menu. (Read more about window shade and other implementations of the Swiss-Army-Knife navigation in Chapter 13.)

FIGURE 5.9: This excellent integrated Tutorial pattern is from the game N.O.V.A.

FIGURE 5.10: This is how the Flipboard app implements the Tutorial pattern using a Watermark.

When the page first loads, it is fully functional as is, with no additional actions needed by the customer. However, it also includes the additional Watermark, which points out the menu ribbon. If the person chooses to ignore the ribbon, well and good—she is free to engage with content instead. However, the overlay is obviously, yet unobtrusively, explaining how to get to "all the good stuff." Voilà—an integrated Watermark Tutorial.

When and Where to Use It

Deciding whether to use the tutorial is easy. If any of the evaluators have a legitimate reason to be confused during user testing, go ahead and put one in. Tutorials are fairly inexpensive to create, can easily solve user frustrations, and help people feel like an expert from the get-go.

Why Use It

Sometimes, as in the game Myst, it's the customer's job to figure out the puzzles. Most of the time, however, for social media and e-commerce apps, decreasing cognitive friction is your duty as a designer. Tutorials help—use them.

Other Uses

Another excellent application of the Tutorial pattern is repeated controls. For example, in the search results, often there are multiple nonobvious functions that the person can take on a given row, and people have trouble figuring out that multiple taps on a single result are even possible. One solution to this problem is to put several strong calls to action on each row. The problem is that this makes the interface busy because repeated controls clutter the interface with chart junk. After the initial discovery is complete and the person learns how to use the app, these controls serve no useful purpose. In this case a tutorial can help educate the customer about the available functionality. After the customer understands how to use the system, she no longer needs all the chart junk that would have otherwise been there to clutter the interface for the rest of the life of the app.

Pet Shop Application

As described in the "Pet Shop Application" portion of the "5.4 Pattern: Welcome Animation" section, Figure 5.11 shows an example sticky-note wireframe of the row-level integrated tutorial. Note the integrated tutorial state and then, after the customer scrolls down or navigates to another page, the "after" state, without the tutorial.

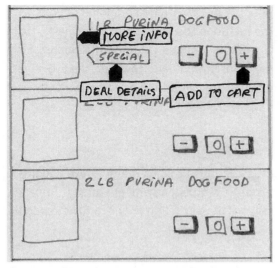

FIGURE 5.11: This wireframe shows a row-level integrated Tutorial in the search results of the Pet Shop app.

Tablet Apps

This pattern applies to tablet apps the same way it does to apps on mobile phones.

! Caution

Don't overdo it. You don't have to explain *all* the functions. Including the bare minimum the person needs to survive your great app design is sufficient. Let people discover the rest on their own.

Related Patterns

13.5 Pattern: Watermark

13.6 Pattern: Swiss-Army-Knife Navigation

5.3 Antipattern: Sign Up/Sign In

CHAPTER 6

Home Screen

See if you can spot a crucial difference between the following two statements:

A. "I heard this great joke on a train today. It was told to me by a gray-headed man while we were riding together on the train from Santa Clara...."

B. "A mushroom walks into a bar. The bartender says to him, 'We don't serve your kind here.' And the mushroom replies, 'What's wrong? I'm a fun-guy!'"

OK, maybe that's not the best joke, but I have a point to make: Even if both statements contain about the same number of words, there is a crucial difference between them: Statement A tells you *about* the story, whereas statement B *actually tells the story*. That is the difference between good and great home screen patterns: The good ones actually tell you the story. The rest, well....

6.1 Pattern: List of Links

Sometimes called Hub-and-Spoke (first documented by the User Experience (UX) w expert Jennifer Tidwell in her essential 2011 book *Designing Interfaces* from O'Reilly), List of Links is a popular and venerable design pattern used all over the mobile world on all manner of platforms and applications. Unfortunately, this pattern frequently tells people *about* the story, rather than telling the story itself. Here's how to spruce it up to improve its usefulness.

How It Works

The home screen acts as a hub that presents a bunch of links or icons of primary functions or popular views that can be obtained with the app.

Example

You don't have to look hard to find an example of this pattern. Travelocity (see Figure 6.1) offers a good one that also highlights the key issue with this pattern: telling customers about the story (instead of the story itself).

FIGURE 6.1: The Travelocity app uses a typical List of Links pattern.

The screen makes it clear what information can be obtained within the app and what actions you can take if you engage with the app as a customer. The icons are clear.

However, despite the clarity of the display, the screen leaves the customer feeling... a little blue (or a little gray, if you are viewing this screenshot in gray scale). There is absolutely no information that pertains to the customer: Everyone gets the same set of links. How would you make this static page tell customers more of a story?

One approach is to use notification badges somewhere on the page or on the individual icons or links. One example of this is the older version of Google Plus (see Figure 6.2), where notifications are featured prominently on the top of the screen (the 3 in the red box).

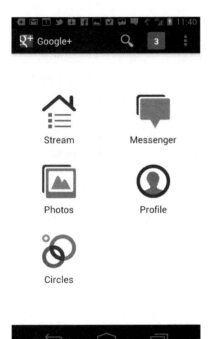

FIGURE 6.2: Google Plus List of Links tells more of the story.

A simple variation of this would be to put notifications as smaller badges on individual icons—for example, place a (1) on the Photos icon if someone shared a new photo and a (2) on the Circles icon when two people have been added to your circles, and so on. The additional notifications would help the List of Links provide a few more details of the story (rather than telling people about the story).

When and Where to Use It

List of Links is the default pattern most people should consider as the first design for the home screen because List Of Links does a great job of cataloging various aspects of the app's functionality. Even if you do not end up ultimately using this pattern for your home screen, drawing a List of Links for your app is a great exercise in organizing the app's Information Architecture (IA) and cataloging possible use cases and those of your competitors.

Other patterns in this chapter are better for certain applications; however, List of Links is the basic default you should consider if your app covers a lot of highly variable functionality. Consider that in the Travelocity example (refer to Figure 6.1) the customer can book hotels, flights, or cars; read about new and exciting vacation destinations; and even find gas. This is a great deal of useful functionality to pack into a mobile app!

Why Use It

The List of Links pattern is one of the easiest and most intuitive for a novice to navigate. It's easy to design and build (provided you have a decent icon designer), and it launches instantly because it does not require a server call to retrieve any information that is not already on the phone. Even if you do decide to go to the server to grab the badges, as shown in the Google Plus app (refer to Figure 6.2), these updates are typically single digits, which require minimal download time.

Other Uses

One popular variation from this basic pattern is the Grouped List of Links shown in the Southwest Airlines app in Figure 6.3.

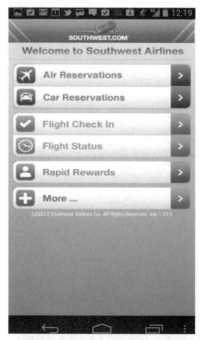

FIGURE 6.3: The Southwest Airlines app shows a grouped variation of the List of Links pattern.

This is a great variation for apps that have a lot of links because it enables a logical grouping that helps reduce the cognitive load on the customer.

Pet Shop Application

List of Links enables designers to succeed with a wonderful variety of IAs. For example, Figure 6.4 shows just two: the IA on the left enables exploration by type of *object* (Dogs, Cats, Birds, Fish, Reptilians, and Other), whereas the IA on the right is more centered on what you can *do* (Find Pets, Care & Feeding, Local News, and Your Profile). It's a great exercise to pause here and imagine a few other alternative IAs that fit the Pet Shop theme. Which one should you choose? Obviously the one on the left emphasizes the e-commerce application, whereas the one on the right is more suited to social networking applications. The one you ultimately choose depends on what your app is designed to do.

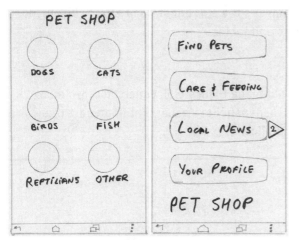

FIGURE 6.4: The List of Links pattern supports multiple IAs.

The badge update mechanism mentioned earlier is also included so that some links (for example, Local News in the right-hand design variation) show the new updates that have come in since the last time that particular area of the app was visited.

Tablet Apps

This pattern always feels a bit dry on the tablet with a large expanse of space available to the viewer. List of Links is less suited to the tablet in general. If you do use List of Links, consider having a split pane view with one of the other patterns in this chapter shown in the other pane.

! Caution

None

Related Patterns

6.4 Pattern: Browse

6.2 Pattern: Dashboard

Sometimes, the app lends itself uniquely to displaying a current state and trends that can be expressed using graphs and tables. Take full advantage of this situation with the Dashboard pattern.

How It Works

When the customer opens the app, he sees the current state of affairs displayed as a dashboard—for example, graphs and tables.

Example

An exceptional mobile dashboard is Mint (see Figure 6.5), which displays an excellent Overview dashboard that shows a snapshot of the current financial state: inflows and outflows, budget, and cash flow.

FIGURE 6.5: The Mint app is an excellent example of the Dashboard pattern.

Note that the graphs and tables are compact, leaving some room for the Alerts and Advice sections, which are prominent and represent a large portion of the value proposition of the app.

When and Where to Use It

Any time you can obtain some numbers that are important to the person using the app, Dashboard is the pattern to use. Financial and banking apps are obvious choices, but elements of this pattern can be pulled in almost anywhere. For example, travel apps can track prices on hotels and flights for recently researched destinations, issuing alerts when it's time to pull the trigger and purchase the trip. Social media can signal how many new friends you've acquired since the last time you used the app and how many photos, updates, and so on you posted as well as any badges you earned. Even the simple examples of using icon badges and notification you saw in "Pattern 6.1: List of Links" also constitute a kind of a primitive dashboard.

Why Use It

You swim in the sea of digital information. That means you consume numbers for breakfast, lunch, dinner, and snack. Providing aggregate information that helps you make sense of your numbers and trends is an increasingly crucial function for which mobile devices are ideal. A small display space forces aggregation; as a result, dashboard-type displays are becoming increasingly common. Last but not least, the best dashboards enable your customers to grok the complete story at a glance, while also making it possible to drill into life-defining questions, such as "*Why* did the mushroom think he was such a fun-guy?"

Other Uses

None

Pet Shop Application

You can imagine several scenarios for the Pet Shop app in which dashboards would be useful. One idea is to track the health and well-being of the pet. The dashboard in Figure 6.6 tracks past and upcoming vet visits, distance walked, and weight of the pet.

If your pet were for some reason in poor health (or just young), you could watch his progress toward recovery and his healthy growth.

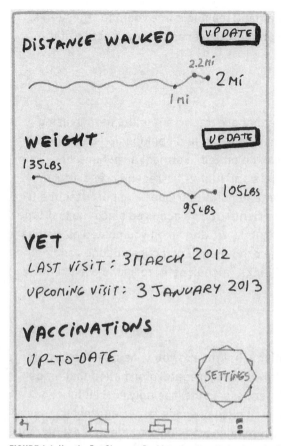

FIGURE 6.6: Use the Pet Shop app Dashboard pattern to keep track of Fido.

Tablet Apps

Dashboards can be fantastic for tablets. Greater screen real-estate allows for extensive multisectioned displays with highly customizable *dashlets* (small graph or table displays) that the customer can rearrange as desired. Tablets are an ideal device to look at dashboards, especially in the context of shared attention, such as office meetings or contextual discussions right on the factory floor. At the present, tablet dashboards are greatly underused—don't let this opportunity pass you by!

! Caution

When creating a dashboard avoid overstuffing it with data. This is easy to do with so much information available right at your fingertips. Test with your customers to make sure your design displays only the most important things upfront and allows people to dig deeper into the information as needed by tapping the desired section.

If one dashboard is good, are multiple dashboards even better? Sometimes, too much of a good thing is...well, bad. Avoid the page carousel effect of multiple dashboards that require your customers to swipe left to right around an endless list of pages. When it comes to seeing more information, *scrolling* is much more intuitive than paging side to side. If you are forced to have more dashboard information than can fit comfortably on one screen, use scrolling instead of side-to-side pagination because scrolling allows for longer pages, which use irregular screen space formed by dashlets more efficiently. For instance, on a long page you can alternate as many large and small elements as needed, instead of forcing one large graph and a few smaller dashlets into a complete page layout on every individual dashboard page. If you must have multiple dashboards, use the Tabs pattern instead (see Chapter 8, "Sorting and Filtering") to identify each dashboard page or use the simple drill-down method, as shown in the Mint example (refer to Figure 6.5).

Related Patterns

6.1 Pattern: List of Links

8.5 Pattern: Tabs

6.3 Pattern: Updates

Whenever you have regular relevant personal information of interest to the customer, consider using the Updates pattern on your home screen.

How It Works

The home screen shows one or more posts or messages from the customer's update stream.

Example

There are many examples of this pattern, from e-mail apps such as Gmail to news apps such as CNN. However, where this pattern really shines is in social media apps such as Twitter and LinkedIn. Perhaps the quintessential and richest example is Facebook (see Figure 6.7).

The Facebook app home screen shows a running summary of updates of various different types, all arranged in the order of Most Recent First. This design is easy to understand and use and immediately tells the freshest story possible anywhere. Note also the roll-away side menu that Chapter 13, "Navigation," covers in more detail.

FIGURE 6.7: The Facebook app's implementation of the Updates home screen pattern is excellent.

When and Where to Use It

Regular updates that are of some interest to the customer are the prerequisite for using this pattern. Thus, it is most often used in communications-driven apps, such as e-mail and social media, and less so in e-commerce, e-readers, and other content- and action-centric applications.

Why Use It

Updates tell the story—pure and simple. Recall the earlier discussion of the Google Plus app in Figure 6.2. The older design of the Google Plus app notifies the customer that he had three updates. It is slightly better than no notifications, but still it tells no story—but only tells *about* the story—and requires the customer to make the extra tap to reveal the details. In sharp contrast, the Updates design pattern makes the story the front centerpiece of the display. For the apps where the story matters, this is an effective pattern because it enables your customers to immerse themselves in the story from the beginning, which means they navigate less and stay engaged longer. Compare the older design of the Google Plus app shown in Figure 6.2 with the recent design that uses the Updates pattern as shown in Figure 1.7 in Chapter 1, "Design for Android: A Case Study."

Other Uses

In addition to personalized feeds from social media and e-mail apps, news stories of general interest are also appropriate for the Updates pattern implementation. Here you have a choice: Whereas typical updates are posted in the order of Most Recent First, news can also be posted in order by Most Popular First or in multiple sections, such as Breaking News, Most Popular Stories of the Day, and Editorials.

Pet Shop Application

Figure 6.8 demonstrates the Updates pattern designed with the screen in two sections: Local Canine News and New Puppies on the Market. Both sections assume you have identified some pieces of information about yourself: where you live and what kind of pet you are looking for. Note that you do not need to label each section explicitly if they have different layouts (though there is nothing wrong with labels). This design saves some screen real estate by not labeling the New Puppies on the Market section because it should be pretty clear to the customer even without the label that it has some "featured" type items that the system judged relevant for him or her. This arrangement also enables you to experiment with this important section and create the perfect mix of content for each customer without being constrained by the section title.

This kind of information might be useful to a dog enthusiast, who would be the target customer for this kind of app. The local news section is shown first because it is less likely that there is much local dog news, so these events are more urgent/important. This section can be entirely optional. For example, if there is no local dog news happening right now, the section might be empty. Verify any assumptions you make in your own designs through user research that will help you understand the needs of your market and design accordingly.

The second section can be based on the last search the customer performed or set up on a separate page, looking for new arrivals on the market of dogs and puppies of a particular breed. There are usually more updates to this section because inventory turns over quickly. Therefore, this section can extend well below the *fold*. (Above the fold is the portion of the screen that is visible without scrolling.) Depending on the design, the entire page can be made scrollable, so the news, once read, can be scrolled off the screen easily. This mixed content updates is a useful model adopted from Facebook; as you can see it can be successfully applied in many different situations.

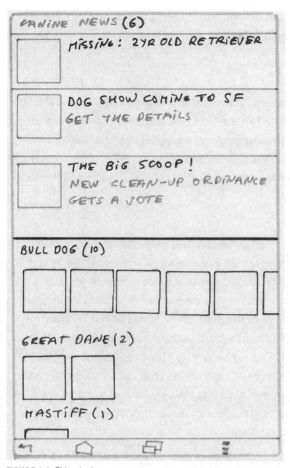

FIGURE 6.8: This wireframe presents an alternative approach to mixed media: the two-sectioned Updates pattern in the Pet Shop app.

Tablet Apps

Updates on the tablets are fantastic and constitute one of the best tablet home screen patterns. Because screen real estate is much less of a problem than on a smaller device, you can devote more space to tell the story with updates without worrying about the mechanism for hiding the navigation; there is usually plenty of space to display navigation alongside the updates.

Whether you use List of Links, Dashboard, or any other pattern described for your tablet home screen, make an effort to also include a section of updates. You'll be glad you did.

! Caution

Keep it simple, especially on tablets. One or two sections of updates are usually plenty; a single section of mixed updates (using the Facebook model) is often best. Remember, the customer is most likely to have the Facebook model in mind when she sees your updates, so she will expect the sort order to be Most Recent First. If you choose a different sort order, make sure that you have a good reason to change, and signal it appropriately to the customer.

Related Patterns

6.1 Pattern: List of Links

6.4 Pattern: Browse

Sometimes, the best home screens are the ones that engage the customer in browsing items or information important to them—provided you supply enough information to avoid pogo-sticking.

How It Works

When the home screen loads, it displays some actual items and item categories of interest to the customer.

Example

A decent example of this pattern is the Amazon.com app (see Figure 6.9). When the app loads, the home screen displays some items of immediate and personal interest to the customer based on his previous search history, or items obtained via the People Who Shopped for X Also Shopped for Y loose-matching algorithm.

Why is this not the best possible example? Even though Amazon.com has a good Browse section, *it is small compared to the overall screen real estate*. There is a lot of branded miscellaneous junk, such as Gold Box, greeting, and navigation. Most important, individual item-level information is completely missing; the customers have no idea why this shoe is shown, what the item's title is, or (this is the key) what the item's price and any associated discount are. All the customer sees is the picture.

A better approach is developed by the redesigned Newegg app, as shown in Figure 6.10.

FIGURE 6.9: The Amazon.com app includes a decent example of the Browse pattern.

FIGURE 6.10: The Newegg app has a better implementation of the Browse pattern.

Most of the page is devoted to products on sale. Each product is shown with a good-sized thumbnail and description, and the price and discount take center-stage. This is an actionable home screen designed so that the customer can immediately make a purchase decision without looking at the detail page. If you scroll past the sale items, you come to the category list, which offers browsing opportunities by category for those customers not immediately interested in sales.

Depending on your app, an even better approach to design a Browse pattern home screen might be to use the 2-D More Like This pattern (similar to Netflix and Gowalla). You can find more information on this useful pattern in Chapter 14, "Tablet Patterns."

When and Where to Use It

Any time you have some content that might be of interest to the customer, Browse is a great pattern to put to work. A Browse section can be small, like it is on the Amazon.com app or implemented as the 2-D More Like This pattern over the entire home screen as it is in the Netflix app.

Why Use It

Much like the individual updates in the Updates pattern, real items of interest *are* the story. Seeing these actual items front and center tells the story and engages the customer immediately.

Other Uses

You could argue that Updates and Browse are similar patterns, and you are, of course, correct. However, there is one crucial difference: Updates are strictly devoted to showing updates from the people you are already connected with or real updates happening in the system. Browse is another facet of the same idea, but it can be expanded greatly to also show related merchandise, upsells, deals, local inventory, and the like. Most successful apps including Amazon.com experiment with a mixture of items in their Browse streams, making a fresh list each time the person visits the home screen. You should, too.

Browsing content for local information and deals is always popular, and it's great to include these things on a mobile device that tracks a customer's immediate location via on-board GPS. Make sure, however, that this content relates in some way to the person's interests and avoid spam and outright ads.

Pet Shop Application

Figure 6.11 shows a simple example of the Browse pattern created with a gallery of new pets available in the customer's area.

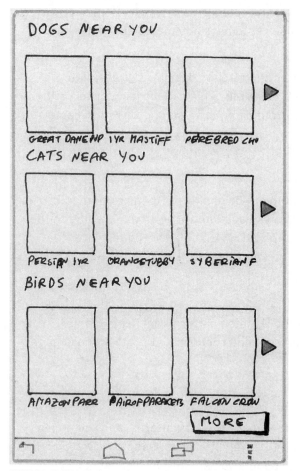

DOGS NEAR YOU

GREAT DANE/MP 1YR MASTIFF   PEREBRED CH

CATS NEAR YOU

PERSIAN 1YR   ORANGETUBBY   SYBERIAN F

BIRDS NEAR YOU

AMAZON PARR   PAIROFPARAKETS   FALCON CROW

MORE

FIGURE 6.11: This wireframe shows a gallery of pets on the home screen of the Pet Shop app using the Browse pattern.

This approach is both effective and attractive as a homepage that enables new customers to know upfront what they can expect from the app. For another great example of a Browse home screen interface design, see the Pet Shop section in the 2-D More Like This pattern in Chapter 14.

Tablet Apps

Browse is practically made for tablets because of the more expansive real estate and free-flowing swiping gestures. Any browse content you can bring into the tablet interface improves the customer's home screen engagement and the appeal of your app.

! Caution

It's easy to yield to the pressure from the marketing department to include items that have nothing to do with the customer and are basically ads. Don't do this. If you do, conversion and customer loyalty will both plummet.

Another thing to watch out for is to ensure that you provide not just images of items but also enough supplemental information so that the customer can make a solid, committed decision to drill down into details to investigate further and possibly purchase or otherwise engage; you want to help the customer avoid pogosticking into multiple items. This commitment on your part must include a large enough thumbnail image for the customer to see the necessary detail and also sufficient supplemental textual information (refer to the earlier Amazon.com and Newegg examples).

Related Patterns

14.5 Pattern: 2-D More Like This

6.5 Pattern: Map

Often the mobile information is highly local and geo-centric. A map-based home screen is the perfect pattern for these applications, provided you get the zoom factor correct.

How It Works

When the home screen loads, it displays a map of the customer's immediate area and shows items of interest.

Example

One great example of this pattern is the Google Maps app (see Figure 6.12).

Maps loads the map of the immediate area, optimized for quick navigation and shows nearby roads and freeways.

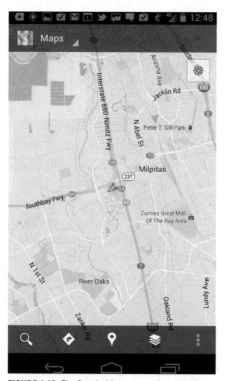

FIGURE 6.12: The Google Maps app makes excellent use of the Map pattern.

When and Where to Use It

Map is a great home screen pattern to use any time geo-centric information is of interest and can be plotted on the map.

Why Use It

Maps are something you are familiar with from an early age. They are intuitive and tell the story. And now, with mobile on-board GPS technology, you have a unique opportunity to display an in-context map that accurately represents the customer's immediate surroundings.

Other Uses

Simple maps for navigation are just the beginning. You can use the Map pattern with any geolocated points of interest. For example, the SitOrSquat app shows nearby restrooms, and Trulia shows homes for sale in the nearby area (see Figure 6.13) .

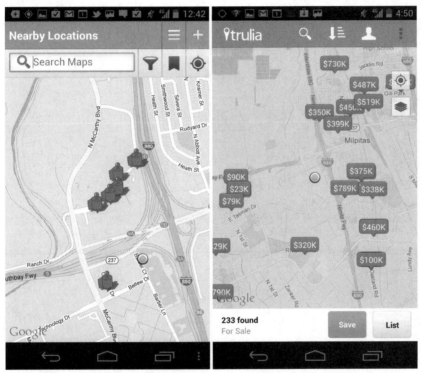

FIGURE 6.13: Compare the Map pattern implementation in the SitOrSquat app (on left) and the Trulia app (on right).

Pet Shop Application

It's easy to imagine showing pets for sale in the immediate area. Figure 6.14 shows a hand-drawn wireframe with a simple implementation of the Map pattern.

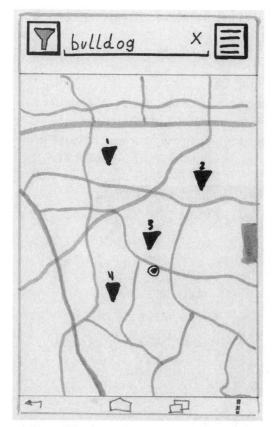

FIGURE 6.14: This is a suggested implementation of the Map pattern in the Pet Shop app home screen.

Tablet Apps

With tablets of all kinds, you must be careful with the GPS data; sometimes it is not available (or not accurate) for tablets that run on a Wi-Fi network. Also be aware that tablets often do not travel as easily or as widely as smaller mobile devices; consequently map-based local information is generally of less interest to tablet users. This is, of course, a generalization. You must do some research to figure out just how pertinent this kind of pattern will be to tablet users of your app.

If you do use the Map pattern on a tablet, one possibility is to show the map corresponding to the *last search* that was done on the device instead of the local map. To see information in a different area, the customer must provide the area of interest as a search parameter, so the tablet app does not need to rely on the GPS data alone for geo-location and can be successfully operated via any Wi-Fi network.

! Caution

One thing to watch out for is the initial zoom level. For example, the initial zoom level in the SitOrSquat app discussed earlier shows a map area that is not sufficiently large to show nearby bathrooms (at least in the heart of the Silicon Valley), as shown in Figure 6.15.

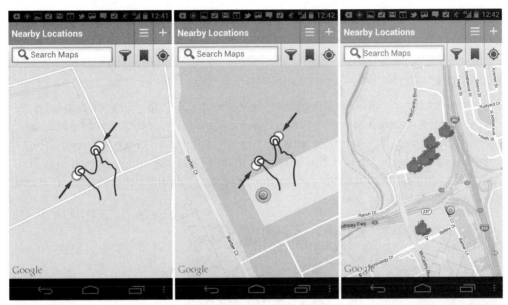

FIGURE 6.15: The initial SitOrSquat Map pattern zooms too close.

It takes several "zoom out" actions to make the nearby bathrooms visible. Inexperienced bathroom (err...app) users might think that there are simply no bathrooms or that the app is malfunctioning. Be sure to zoom out your initial view sufficiently to capture two or more points of interest. In general, it is easier to zoom into a selected area than to zoom out. Zooming in can be done one-handed using a double-tap shortcut, whereas zooming out requires the pinching multitouch gesture that requires two hands (one hand to hold the device and the other hand to pinch the screen).

Related Patterns

8.4 Pattern: Parallel Architecture

8.5 Pattern: Tabs

9.6 Pattern: Local Results

6.6 Pattern: History

Whenever you encounter app engagements that span multiple sessions, it is a great idea to re-engage customers by reminding them of the subjects of previous searches.

How It Works

The home screen displays a list of links or items that represent recent previous queries, with the most recent query listed first.

Example

One of the best examples of this pattern is the Android Global Search app (see Figure 6.16). It shows a mixture of apps, web pages, and contacts that have been recently searched. If customers look for the same "person, place, or thing," they do not need to search again—the information and call to action is right there. The page also tells a complete story about recent activity on the device (versus telling the customer *about* the story).

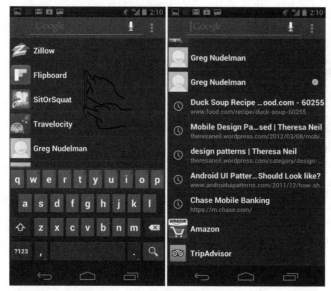

FIGURE 6.16: The Android Global Search app has an excellent History pattern homepage.

When and Where to Use It

Anytime the person looks for the same thing multiple times and needs to be re-engaged in the search process, or when the customer needs to be inspired or reminded of the previous searches, History is the module to use.

Why Use It

Why not? Despite easy availability of search history and its tremendous usefulness, few apps take full advantage of this pattern. For example, the Priceline homepage, as shown in Figure 6.17, simply displays the search screen and offers an option to look in a nearby city (Fremont, CA) or in a different location (marked by Choose a City).

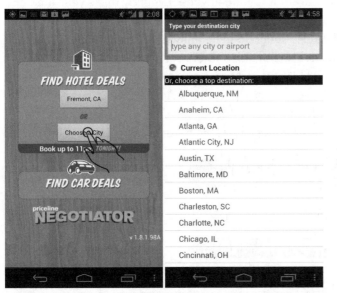

FIGURE 6.17: The disappointing Priceline app homepage doesn't include the History pattern.

The Choose a City function is disappointing, to say the least; it simply displays an alphabetical list of popular destinations. This page would certainly be enhanced by a modest History function. If only the app remembered the last four to seven places you've looked at! Often, booking a hotel is a multistep process, so the customer is likely to revisit the app such as the Priceline Negotiator multiple times in between getting driving directions, calling a friend, texting, looking at Twitter and Facebook updates, looking for a place to eat on Yelp, and so on.

But multitasking only scratches the surface. Recently I had an experience looking for a hotel in the middle of the large metropolis of Los Angeles, which, as it turns out, is actually composed of multiple small towns. Each town has its own hotel inventory, each of which requires a different search by city name. Jumping between searches for Beverly Hills, West Hollywood, and Santa Monica with the goal of finding a reasonably priced hotel at the intersection of all three cities was *most* tedious. Having a basic History module with 5 to 10 recent destinations on the homepage would take care of this common problem.

Other Uses

Search history often acts as a wish-list functionality when it kicks in automatically. Unfortunately, few apps take full advantage of this pattern. For example, the Trulia app forces the customer to tap the Save button to save the search, which in turn, forces him to log in (see Figure 6.18) . This is a lot like my HP printer saying, "You can no longer print in black and white. The magenta ink cartridge has expired." Seriously Hewlett Packard, you are not fooling anyone: Even my three year old knows that you don't need the magenta crayon to draw a black and white picture. You only need one crayon: *black*! Although logging in by itself is not a huge deal, it is completely unnecessary to save a short history of recent searches. Every native app has some local storage space it can use to store history (or a temp guest session token if you prefer to save recent history on the server). Having to tap the Save button to save a search is both tedious and unnecessary; the extra tap is one more thing for your customers to do, and it interrupts the natural flow of searching.

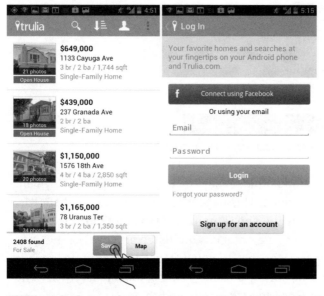

FIGURE 6.18: The Trulia app forces an explicit save and login instead of adding to on-board app History.

A better implementation would be to provide a search history function that automatically remembers the last 10 to 15 searches run by the customer. People looking for a home tend to run the same few queries over and over again, so automatically remembered searches would serve people well in this context. Do the customers need to log in at all? Only if they tap a Share This Search or Share This Property button that would replace the Save button to share information with other people or multiple devices. At this point, logging in would be both expected and natural.

Pet Shop Application

The History pattern is perfect for pet searches because the shopper might be looking at several different breeds while dealing with a rapidly changing local inventory. For example, if a customer looks for guard dog puppies, the History module would be advantageous; he could redo local searches for "Dogue de Bordeaux," "Bouvier des Flandres," and "Boerboel" without twisting his fingers while trying to retype these names. (See Chapter 7, "Search," for more ideas about helping customers with typing long or difficult queries.) Figure 6.19 shows a simple wireframe of how this History-forward page could look, augmented with the Updates pattern, which shows the count of new dog inventory within 20 miles of the area from the last time the search was run.

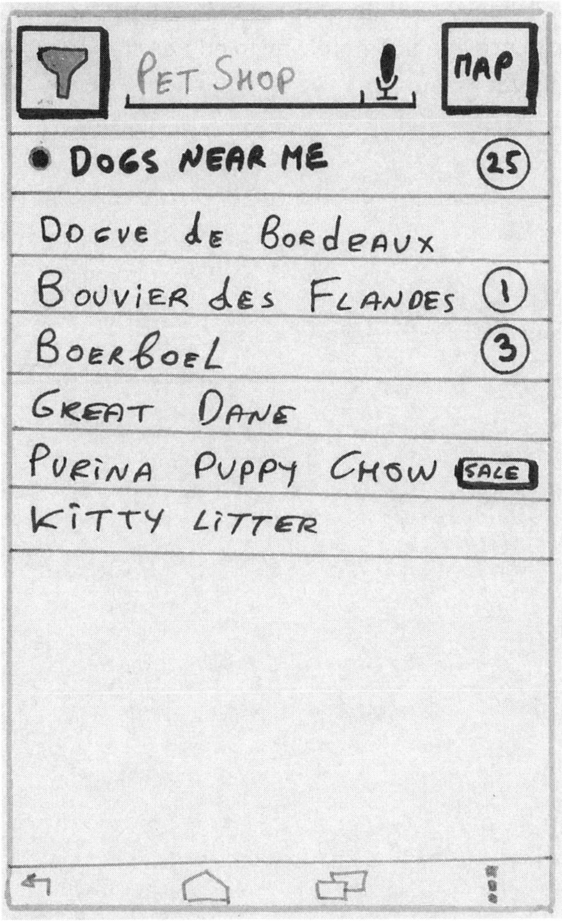

FIGURE 6.19: The Pet Shop app History pattern home screen with local Updates is highly effective.

For Pet Shop customers, this kind of home screen would be "killer functionality," and might just be the selling point for the entire app.

Tablet Apps

Typing on tablets is less tedious, and connections are often faster than on other devices. That does not mean that people using tablets like to think any more or any harder than the people using smaller mobile devices. History is a great module to include in everything from a simple map app to sophisticated shopping services.

! Caution

No matter the detractions, having the History module is almost always better than not having one. That said, remember to provide a simple way to edit or clear the history because it can become a big privacy issue.

Related Patterns

5.3 Antipattern: Sign Up/Sign In

6.3 Pattern: Updates

7.2 Pattern: Auto-Complete and Auto-Suggest

7.3 Pattern: Tap-Ahead

CHAPTER 7

Search

Search is a fundamental mobile activity. Think about it—mobile is much less about creating stuff (unless you are talking about taking pictures or writing an occasional tweet). Instead, you use mobile devices mostly for finding stuff. Riffing on Douglas Adams' *Hitchhiker's Guide to the Galaxy*, mobile devices help you find places to eat lunch, people to eat lunch with, and directions to get to the restaurant, which helps everyone to get there sometime before the Universe ends— which makes search patterns important.

7.1 Pattern: Voice Search

Audio query inputted via an on-board microphone is used as input for searching instead of a keyword query. Typing on the phone is awkward and prone to errors. This makes audio input a great alternative to text.

How It Works

Usually, the searcher taps a microphone icon, causing the device to go into *listening mode.* The searcher speaks the query into the on-board microphone. The device listens for a pause in the audio stream, which the device interprets as the end of the query. At this point the audio input is captured and transcribed into a keyword query, which is used to run the search. The transcribed keyword query and search results are shown to the searcher.

Example

One of the most straightforward implementations of the Voice Search pattern is the standard input box for writing text, augmented with a microphone icon, as exemplified in Google's native Android search. (See Figure 7.1.)

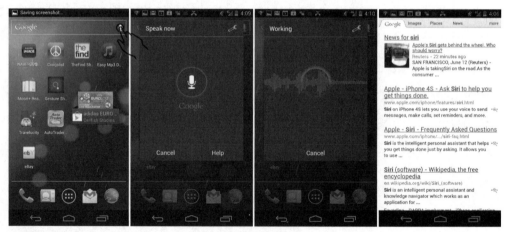

FIGURE 7.1: Google's native Voice Search in Android 4.0 is straightforward.

When and Where to Use It

Most apps that have a search box can also use the Voice Search pattern. For example, the Yelp app, as shown on the left in Figure 7.2, does not currently include the Voice Search feature, but it can be easily augmented with a microphone icon, as shown in the wireframe on the left.

FIGURE 7.2: Adding the Voice Search pattern to the Yelp app would be easy from the UI standpoint.

People often use Yelp while they're walking around with a bunch of friends and talking about where to go next. In this case, simple voice entry augmentation makes perfect sense: Speak the query into the search box (which is quite a natural behavior as part of the human-to-human conversation already taking place) and share the results with your friends by showing them your phone. Then, after the group decision has been made, tap Directions, and use the map to navigate to the place of interest.

Why Use It

Most mobile search is done "on the go" and in context. Given how hard text is to enter into a typical mobile phone (and how generally error-prone such text entry is) voice input is an excellent alternative. Other important considerations for using the Voice Search pattern are multitasking activities such as driving. Driving is an ideal activity for voice input because the environment is fairly quiet (unless you are driving a convertible), and the driver's attention is focused on a different task, so traditional text entry can be qualified, to put it mildly, as "generally undesirable."

Other Uses

The release of Siri for the iPhone 4S kicked into high gear a long-standing race to create an all-in-one voice-activated virtual assistant. Prior to Siri, Google had long been leading the race with Google Search: the federated search app that searched across phone's apps, contacts, and the web at large. Vlingo and many other apps took the Voice Search pattern a step further by offering voice recognition features that enabled the customer to send text messages or e-mails and do other tasks by simply speaking the task into the phone. However, none of the apps have come close to the importance and popularity of Siri. Why? There are many reasons, including the mature interactive talk-back feature in Siri that enables voice-driven question-and-answer interactivity, including the amazing capability to handle x-rated and gray-area questions with consistent poise and humor, as shown in Figure 7.3 (in other words, Siri has something of a personality). Another important feature was a dedicated hardware Siri button (on iPhone 4S you push and hold the Home button to talk to Siri) that enabled one-touch interaction with a virtual assistant without having to unlock the phone.

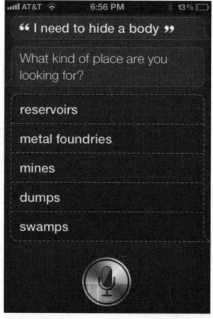

FIGURE 7.3: Siri responds to a Voice Search query: "I need to hide a body."

Although it's pure speculation at this point, one of the applications of Google's voice recognition technology could be the same sort of virtual assistant for your phone or tablet, activated by pressing (or holding) one of the hardware buttons (the Home button would be a good choice). Added security can be achieved via voice-print pattern recognition. Voice recognition technology would also help distinguish your voice patterns from those of other people in loud, crowded places, thereby further increasing the personalization of the device and making it completely indispensable (if that is even possible at this point!).

If this becomes the case, dedicated in-app Voice Search (refer to Yelp in Figure 7.2) can be completely superceded by the Google virtual assistant. For example, the customer could say, "Assistant: search Yelp for xyz." The assistant program would then translate the voice query into keywords using advanced personalized voice recognition, open the Yelp app, populate the search box with the keyword query, and execute the search.

In some Google Search apps, the simple action of bringing the phone to your ear forces the app into a listening mode by using input from the on-board accelerometer to recognize this distinctive hand gesture. Unfortunately, this feature does not seem to be automatically enabled on Android 4.0 as of this writing. It is, however, an excellent feature and one that should come included with the voice recognition because it makes use of what we already do naturally and without thinking, so the design "dissolves in behavior."

The role of voice input is not limited to search. It can be used for data entry and basic tasks as well. For example, while driving you could push the button and say, "Text XYZ to James," and the device will obey. I should also mention that Google is not the only supplier of voice recognition technology. For example, Nuance communications, the maker of Dragon Naturally Speaking products, is likely the largest and most vocal (pardon the pun) distributor of speech-recognition software. As of this writing, the Target app uses technology licensed from Nuance for its voice recognition feature.

Pet Shop Application

Just as in the earlier Yelp example, you can use voice recognition to search for a specific pet. The customer would launch the Pet Shop app and then swing the phone up to his ear and speak a search query, such as "black lab." When the customer has a pause in speech, pushes the Done button, or simply swings the phone down, the query activates and displays the appropriate search results.

Tablet Apps

For Voice Search, tablets are different from phones. Although there is some debate about this (and no official studies have yet been performed) anecdotal evidence points to typing on the tablet being not quite as challenging as it is on the phone. Thus voice input for tablets is likely to be more error prone. While using a tablet, the person is also less likely to be multitasking in a loud environment or be engaged in an activity that requires the user's attention to be placed outside the visual interface of the device (driving, for example)—most of the tablet use happens at home or work. Does this mean Voice Search is not useful on the tablet? Not at all. There still exists an opportunity for high-end, high-touch, visual interaction with a virtual assistant software program. Apple's original vision for the tablet device, The Knowledge Navigator, created in 1987 (sorry, Google, you were not yet born at that time) involved exactly that kind of speech recognition interaction with the device.

The best way to implement a high-end, personalized virtual assistant might be to create a *hybrid* of software plus human virtual assistant. The person using the tablet would get high-end service with a consistent, pleasing visual and auditory representation. Given Google's reputation for awesome inventive geekiness, highly customized animated Obi One, Jarvis, and HAL virtual assistants (as well as various Playboy models, anime characters, and maybe a little something for the millions of John Norman fans) complete with high-end graphics and voice simulations might be coming soon to the Android tablet near you. Perhaps this book can serve as an inspiration?

! Caution

Voice recognition is still a fairly new technology, and despite the apparent similarity of the interface, there are many important considerations and ways to get this pattern wrong:

- **Don't forget the headset:** Some users of the technology will be on a Bluetooth or wired headset. Ideally, Voice Search can be activated by using the buttons on the headset, without having to touch the phone. For example, with Apple's Siri: "When you're using headphones with a remote and microphone, you can press and hold the center button to talk to Siri. With a Bluetooth headset, press and hold the call button to bring up Siri." (See http://www.apple.com/iphone/features/siri-faq.html.) Similar convenience features are conspicuously absent from the Android 4 interface for the simple reason that the headsets from various manufacturers lack consistency in the hardware configuration. (In other words, *there is* no "center button.") As mentioned earlier, this needs to change. Convenience is the key for Voice Search on Android to be a contender.

- **It's not "Done" until the fat finger sings:** The Android 4 implementation of the Google's native search (refer to Figure 7.1) waits for you to stop talking before accepting the voice query. This works most of the time, but it can be a serious problem in loud environments, in which the interface fails to stop and keeps listening for almost a full minute! Always remember to provide a Done button to stop the input. One of the best implementations is to make the microphone icon act as a Done button. Of course, it must also look "tappable."

- **Extremely loud and incredibly personal:** In loud environments in which other people are talking, it's hard to parse the owner's voice from the background of other conversations. Fortunately, voice imprint is as unique as your fingerprints, and with some "training" the user's unique vocal patterns can be parsed out of the background conversations in the crowd. Voice imprint has a lot of privacy issues.

- **Full-circle audio experience:** The Driving Mode paradigm that exists in certain older Android phones is an antipattern. There is an excellent reason no one uses vi editor for coding Java or writing books. As Alan Cooper so eloquently stated in *About Face* (2007, Wiley), switching modes is tedious and error-prone, not to mention downright dangerous while driving. The only reason why Airplane mode works is because a nice flight attendant tells us it is time to turn it on. For all other applications, the system simply must make an effort to match the output mode to the user-selected input mode. For example, if Yelp were asked for directions to a museum using voice input, chances are the customer is doing it while driving (the app can also detect when someone is moving at car speed automatically using the on-board GPS). This means that the output directions should also be available using voice. Ideally, Yelp should read out loud step-by-step driving directions if the customer asks for them via a simple voice command such as Tell Me or Give Me Directions. This completes the 360-degree, full-circle Voice Search experience. As of this writing, for example, Apple's Siri has made inroads into integrated audio directions—a feature Android sorely needs to compete in the Voice Search space. Of course, this also works great for folks with disabilities.

- **Watch out for uncanny valley:** *Uncanny valley* (http://en.wikipedia.org/wiki/Uncanny_valley) is a term coined by Masahiro Mori to describe the strong revulsion people feel for robots that are almost, but not quite, human in appearance. One of the consequences of the uncanny valley is that *more realism can lead to less positive reactions*. As virtual assistants improve, and especially when they acquire visual appearance as well as voice, you need to make sure you make the purely digital entities look decidedly less than human. Uncanny valley might turn out to be an especially dangerous place for high-end hybrid human-software assistants. In their brilliant book *Make It So* (Rosenfeld Media, 2012), Nathan Shedroff and Christopher Noessel mention another issue: Authentically human-appearing digital assistants increase the expectations of the near-human capabilities, which sharply increases the person's level of annoyance when the assistant screws up the task or fails to understand the person correctly. They suggest an elegant way of solving this problem by making a digital assistant into a talking pet: Although most owners consider their dogs intelligent, a talking dog would engender both lower expectations and the higher wow factor, while avoiding the uncanny valley entirely. Here's my own recommendation: Using mid- and low-fidelity animation for pure or hybrid digital assistants (think animated Obi Wan Kenobi from the series Clone Wars or the LEGO Star Wars games, instead of video of a live actor) should likewise be helpful in achieving the same goals.

Related Patterns

13.5 Pattern: Watermark

7.2 Pattern: Auto-Complete and Auto-Suggest

Auto-Complete and its sister pattern, Auto-Suggest, are broad classifications of keyword-entry helper patterns. Both reduce the number of characters the person needs to type and reduce the number of entry errors and queries that produce too many or too few results.

How It Works

When the person enters one or more characters into the search field, the system shows an additional "suggestions layer" that contains one or more possible keyword combinations that in some way correspond to what the person has entered. At any point, the person has the option to keep typing or select one of the system suggestions.

Strictly speaking, Auto-Complete uses a part of the query the person typed in as a seed to providing suggestions (so that the suggestions include the original keyword or fragment). This does not always work perfectly on a mobile device because many times a small fragment contains fat-fingered misspellings. That's where Auto-Suggest comes in.

Auto-Suggest has more "freedom of movement" than Auto-Complete, providing keywords and queries that include

- Spelling corrections
- Controlled vocabulary keyword substitutions
- Synonyms of what the person originally typed in, query expansions, and so on

The suggestions work best when they are a clever combination of Auto-Suggest and Auto-Complete, with the system drawing the best ideas from multiple sources.

Example

Google Android search is a great example of the combined pattern, splitting the suggestions layer into two sections: first providing three auto-complete ideas and then auto-suggesting some contacts and apps that can be found on the phone (see Figure 7.4).

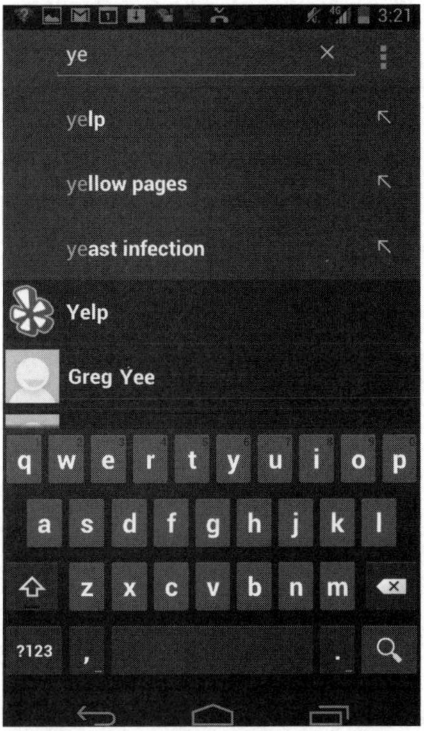

FIGURE 7.4: The Auto-Complete and Auto-Suggest patterns work in tandem in Android 4.0 Native Search.

When and Where to Use It

Any time there is a keyword query entry box, Auto-Suggest and Auto-Complete are both great patterns to implement. As search expert Marti Hearst reports in her book, *Search User Interfaces* (Cambridge University Press, 2009) These features generally rate great on usability and work well with other user interface (UI) patterns.

Why Use It

For most people, typing—especially on the mobile device—is tedious and prone to errors. Generally, the less typing you do on the phone, the better. Therefore, any UX pattern that can assist a person in entering information is a big win.

Auto-Complete and Auto-Suggest help reduce errors and increase satisfaction in multiple ways:

- **Reduce spelling errors:** By reducing the total keys that need to be pressed, the system reduces errors associated with simply fat-fingering incorrect keys.

- **Improve query specificity:** If the suggestion includes more keywords than the user was originally planning to enter, the customer often picks the more specific query, which ultimately makes her happier by providing her with the inspiration to enter "Nike Shoes" instead of just "Nike."

- **Reduce zero results:** Often queries are spelled correctly but include incorrect or conflicting keywords, which produce bad results or no results. Giving the person appropriate suggestions before she finishes entering the entire query often allows the system to forestall zero-results conditions before they occur. If the person starts typing **Harry** and the system displays a suggestion "Harry Potter and the Deathly Hollows," the person is less likely to paint herself into a corner with an incorrect query such as "Harry Potter and the *Sleepy* Hollows."

Other Uses

Auto-Complete and Auto-Suggest can draw from many other resources to improve the quality of the suggestions:

- **Local:** Mobile phone use cases are unique, because these devices are used literally "on the move." Thus auto-suggestions must take into account local results (obtained via on-board GPS or wireless signal triangulation) whenever possible. For example, depending on the app and use case, a query such as "Coffee" could easily include auto-suggestions such as "French Roast Coffee" for online purchases and one or two nearby coffee shops.

- **History:** The Auto-Suggest pattern need not always make use of the Internet connection. One of the most important mobile User Experience (UX) patterns is *Re-engagement*, which means picking up the previous task after an interruption of some sort (phone call, text, needing to find directions, and so on). Therefore, one of the most important functions of Auto-Suggest is to recall previous queries (History) that can be stored locally on the device using the on-board app databases (refer to the Browse and History patterns in Chapter 6).

- **Voice Search:** At the time of this writing, voice queries typically do not produce auto-suggestions. This is surprising because voice recognition is often worse than the fat-finger typing recognition. Voice query Auto-Suggest is one interesting application to look into, particularly for nondriving applications in which the system needs to display only the suggestions on the screen rather than reading them back to the user.

- **Jump into other apps:** One common use case, especially with heavily networked apps, is the need to open another app to accomplish a task. Auto-Suggest can help providing a one-touch solution to shorten the task considerably. One example could be a "gas station" query on Yelp. One auto-suggestion could be "Directions to the nearest gas station"—a highly relevant use case when you are about to run out of gas. Hitting this auto-suggestion would jump directly into the Google Maps app to display directions. This kind of suggestion is also highly relevant to Voice Search use cases because it could provide a full-circle voice interaction with the system reading off directions to complete the experience. Other ideas could include providing suggestions that jump directly into the MP3 player if the query is "Like a Rolling Stone," or into a book reader to see a sample of *War and Peace*. Note that it helps the customer to understand what will happen if auto-suggestions that jump directly into other apps include an icon of that app somewhere in the auto-suggestion row.

Pet Shop Application

With the variety of common names for dog breeds (and difficulty of spelling them) it's easy to envision a useful combination of the Auto-Suggest and Auto-Complete layer for the Pet Shop app, as shown in Figure 7.5.

In this simple example, the person types in **Mas**, and the suggestions layer presents the Auto-Complete options Massive and Mastiff as possible query completions, thereby forestalling the common misspelling Mastif, which would have likely resulted in zero results. In the same suggestions layer, Auto-Suggest also kicks in with English Mastiff, Neapolitan Mastiff, and an interesting keyword variation Bullmastiff, a popular Mastiff breed that the person may not have thought of using as a query.

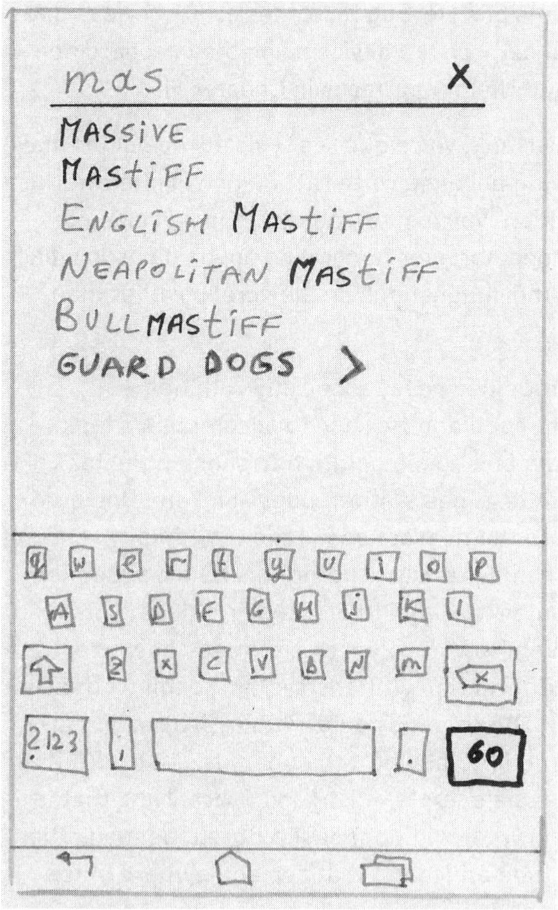

FIGURE 7.5: This wireframe is for a useful combination of Auto-Complete and Auto-Suggest in the Pet Shop App.

Mastiff is also a generally accepted synonym for a query "large guard dog," so the auto-suggest layer can expand the original query by suggesting a category Guard Dogs, which can expand into a number of related breeds the person might not have thought of originally, such as Doberman, Rottweiler, American Bulldog, and so on. Both Auto-Suggest and Auto-Complete automatically scope the suggestions using a controlled vocabulary with a preset list of recommended search terms that match common tasks the app supports.

Tablet Apps

Tablet auto-suggestions represent a different use case from auto-suggestions on mobile devices. In principle, large tablets do support mobile activities; in practice, the mobility pattern for a typical consumer large tablet device is found in the area between the refrigerator and the couch, as user researcher Marijke Rijsberman explains in her perspective "A Fine Line: The iPad As a Portable Device," which is in my first book *Designing Search* (Wiley, 2011). Simply put, it is more common for large tablets to be used as casual, "lean back" devices.

Typing on large tablets is easier and less error prone, so they are closer to desktops and can use the same auto-suggestion database as a desktop web application. Also, one-tap auto-suggestions that jump directly to a different app are not as important on large tablets as they are on mobile devices because people on tablets are typically not in as much of a hurry and are less likely to mind a few additional taps, as long as it's clear that they are progressing toward their goal. Also, local results are generally not as important as they are on mobile devices; however, they should definitely be included.

Note that this does not necessarily apply to mid-size 7-inch tablets and note-tablet hybrids (refer to Chapter 3, "Android Fragmentation"). These smaller tablet devices are at once more mobile and harder to type on than their large counterparts. For the purposes of this pattern, these smaller tablet devices can be treated as mobile phones, and you should design for them accordingly.

Finally, another consideration is the interface element. In mobile devices, the auto-suggestions layer often occupies the entire page, whereas on a tablet auto-suggestions are presented in a popover layer occupying only a small part of the screen. (For more on tablet design patterns, see Chapter 14, "Tablet Patterns.")

! Caution

If you do provide a custom auto-suggest layer (which is highly recommended) remember to turn off the device's auto-suggest feature.

Remember that mobile phones are a different class of device. They may require a completely different auto-suggest approach (one of such mobile-only approaches is described in the next pattern 7.3 "Tap-Ahead"). Mobile Auto-Suggestions are prioritized differently because they are meant to respond to different needs. Mobile devices need to give higher weight to auto-suggestions based on on-board sensors that are only available on mobile devices. For example, local auto-suggestions, previous mobile search history and category browsing (for example, Guard Dogs, as described in the Pet Shop example) need to be higher on the list than typical desktop web auto-suggest options, which are mainly controlled vocabulary substitutions.

People misspell things differently on desktop web and tablets with full keyboards than on smaller mobile devices. Mobile misspellings mainly arise due to fat-fingering, not from common spelling misconceptions. This dictates ideally using and maintaining a different database for mobile auto-corrections that take the unique nature of mobile keyboards into account.

Related Patterns

7.3 Pattern: Tap-Ahead

7.1 Pattern: Voice Search

7.3 Pattern: Tap-Ahead

Tap-Ahead implements auto-suggest one word at a time, through step-wise refinement, creating a kind of *keyword browsing*.

How It Works

Instead of trying to guess the entire query the customer is trying to type at the outset and offer the best one-shot replacement the way desktop web does, Tap-Ahead on mobile devices guides the auto-suggest interface through the guessing process one phrase or keyword at a time.

This is how it works: When the searcher enters a few characters, the auto-suggest function offers a few query suggestions. At this point the searcher has two choices:

- Tap the query if it is a sufficiently good match for what she is looking for.

- Tap the diagonal arrow on the right side of the screen to populate the search box with the query keywords, and execute the auto-suggest function again.

By giving the searcher the ability to "build" the query instead of typing it, the interface offers a much more natural, flexible, and robust auto-suggest method that's optimized to solve low bandwidth and fat-finger issues people experience on mobile devices. Using the Tap-Ahead interface, customers can quickly access thousands of popular search term combinations by typing just a few initial characters.

Example

An excellent example of this pattern is the Android native search (see Figure 7.6). As you can see from the following example, the Tap-Ahead pattern offers an excellent alternative to typing longer multi-keyword queries.

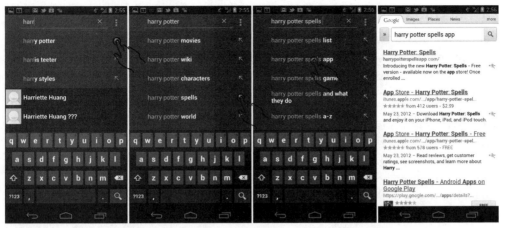

FIGURE 7.6: The Android 4.0 Native Search includes an implementation of the Tap-Ahead pattern.

In this case, by tapping the diagonal Tap-Ahead arrow, the searcher could enter a complex query "Harry Potter spells app" by typing only four initial characters (**harr**) and tapping the diagonal arrow two times. The traditional one-shot auto-suggest interface is unlikely to be able to offer this entire fairly unusual phrase as an auto-suggestion, so the customer is likely to have to type most, if not all, of the 23 characters of the query Harry Potter Spells app.

When and Where to Use It

Use the Tap-Ahead pattern anywhere the auto-suggest is used outside a one-shot controlled vocabulary auto-suggestion and where longer, multistep, multi-keyword queries offer an advantage and create a better set of results.

Why Use It

In contrast to desktop web search, auto-suggest on mobile devices is subject to two unique limitations: It's harder to type on a mobile device and signal strength is unreliable. Tap-Ahead solves both issues in an elegant, minimalist, and authenti-cally mobile way. Tap-Ahead enables the mobile auto-suggest interface to maintain flow and increase speed and responsiveness on tiny screens that is simply not pos-sible to currently achieve with the traditional one-shot auto-suggestion interface.

Is there evidence of this? The author's field research shows that in mobile environments people often select search suggestions they do not need, just to save typing in a few characters. (Read more about this in "Mobile Auto-Suggest on Steroids: Tap-Ahead Design Pattern," *Smashing Magazine*, April 27th, 2011, http://www.smashingmagazine.com/2011/04/27/tap-ahead-design-pattern-mobile-auto-suggest-on-steroids/). Tap ahead effectively resolves this issue.

Other Uses

For the few years that the Android platform has been around, the keyword suggestions have evolved from being an exact match to Google's web suggestions to being its own mobile-specific set. Yet you can do even better in your own app by using a simple trick: Offer Tap-Ahead *one keyword at a time*.

The advantage of the one-word-at-a-time Tap-Ahead refinement interface is that the refinement keywords can be loaded asynchronously for each of the 10 auto-suggestions while the customer makes the selection of the first keyword. Given that most queries are between two and three keywords long, and each successive auto-suggest layer offers 10 additional keyword suggestions, Tap-Ahead with step-wise refinement enables customers to reach between 100 (10 * 10) and 1,000 (10 * 10 * 10) of the top keywords through typing only a few initial characters.

Anecdotally, although Tap-Ahead is useful, few people have discovered its power to cut through tediousness and all the fat-finger mistakes associated with typing. By offering keywords one at a time, the interface is optimized for the Tap-Ahead pattern, so discovery should increase, thereby also increasing the satisfaction. Tap-Ahead one word at a time is an excellent variation of the Tap-Ahead for e-commerce apps.

Pet Shop Application

It's easy to imagine Tap-Ahead being useful in entering complex keyword queries. However, it's not as important with dog breeds, for example, which form a controlled vocabulary. There is scant advantage to provide a Tap-Ahead expansion from Mas to Mastiff to Neapolitan Mastiff because there are not many queries that start with Mastiff. Instead, a simple, traditional one-shot controlled vocabulary auto-suggestion (Mas directly to Neapolitan Mastiff) is a more useful approach because it not only allows the user to pick up standard keyword queries such as English Mastiff and Neapolitan Mastiff but also an interesting keyword variation Bullmastiff and category expansion Guard Dogs (see the "7.2 Pattern: Auto-Complete and Auto-Suggest" section).

Tablet Apps

The owners of large tablets are generally more willing to type a longer query, and low bandwidth is usually less of a problem for them (many tablets are used with Wi-Fi only). Nevertheless, Tap-Ahead is no less useful on tablets, where less work is perceived as a good thing and tapping a suggestion is as easy as tapping the next character on the touch keyboard. There is also early evidence that tablet queries are slightly longer, which also speaks in favor of keyword browsing.

! Caution

The best auto-suggestions on a mobile device come from a database that's different and distinct from the web auto-suggestions database. This is especially true for Tap-Ahead implemented one keyword at a time—but that's how important this function is to creating an excellent search experience!

At this point it's not clear who, if anyone, holds a patent on this functionality. Google began using it first in its general device search and Google App for iPhone; although it is not used for *single* keyword browsing as of the time of this writing. Microsoft and Apple are both likely actively pursuing similar patents.

Related Patterns

7.2 Pattern: Auto-Suggest and Auto-Complete

7.4 Pattern: Pull to Refresh

Search results are refreshed when the customer swipes down (pulls down) on the results. Slick and convenient, this is a great pattern to refresh results that update frequently.

How It Works

The customer is presented with a long list of updates, typically sorted by Time: Most Recent First. The customer typically reviews the list of updates starting at the top, reading the most-recent messages first. When the customer wants to load newer updates, he pulls down on the results list, performing a scroll-up function. Typically, a watermark appears that lets the customer know that when he pulls down and then releases the list, it will update. The system issues an update call, which is reflected by a visible timer, followed by loading of the updated results.

Example

A great example of this pattern is the original application that helped popularize it: the Twitter mobile app. (See Figure 7.7.)

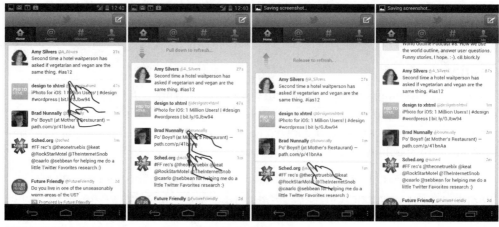

FIGURE 7.7: The Pull to Refresh pattern was popularized in the Twitter app.

When and Where to Use It

Use Pull to Refresh for long lists of search results or updates sorted by Time: Most Recent First. This pattern is especially useful for social update streams, active inboxes, and other long lists that update frequently.

Why Use It

The Pull to Refresh pattern uses a gesture instead of a button, which is always an excellent idea if you can communicate the needed gesture in an obvious and unobtrusive way. For Pull to Refresh, the gesture needed is the one the customer already uses to scroll the results up, so the call to action naturally "dissolves in behavior."

When the customer first loads the results, he typically engages with the list by scanning or reading the newest updates or search results first, starting at the top, and scrolling down the list to read or scan more. When the customer reads far enough down and wants fresher results, he naturally scrolls to the top and keeps scrolling until he reaches the top of the currently loaded results and scrolls past the top of the list. At that point he sees the watermark telling him what to do to load the newest results. This often happens naturally and in the state of flow, when the customer flicks rapidly to scroll the results quickly.

One other point makes this pattern feel natural. The action to pull down on the list "pulls" new data from the server, which is an excellent fit to the customer's existing mental model. This is a fine example of using unique capabilities of mobile and tablet touch devices to expand on the desktop web model of buttons and links.

Other Uses

Most applications of this pattern deal with search results or updates sorted by Time: Most Recent First. Another possible application might be triggered by transversing space instead of time. For example, if your customer looks for points of interest around him as he moves through a city, he has a different set of attractions within walking distance as he moves. Depending on the specific goal of the interaction, you can use the Pull to Refresh pattern to show the search results list sorted by Distance: Nearest First. This way, as the customer moves through the city, he can have an updated list of points of interest around him with a flick of a finger.

Pet Shop Application

One possible way to use Pull to Refresh in the Pet Shop app is to show updates of lost pets. If your pet is lost, for example, you can stay on top of the search with the updates page that tracks found pets in your neighborhood by periodically pulling to refresh the list. However, forcing the customer to do this may be a stressful activity if the list keeps coming up empty or static. If the list is mostly static, instead consider using some sort of a push alert (an alert that is loaded on the device and shown automatically, as opposed to being triggered by some action the customer explicitly needs to take) that notifies the customer when a new pet is found. To create a push alert a polling technology is frequently used, but from the standpoint of the customer, the alert is being "pushed" to him.

Tablet Apps

Pull to Refresh works just as well on medium- and large-size tablets as it does on mobile phones. The vertical space needed to communicate Pull to Refresh should grow proportionally to the size of the device and the extent of a gesture needed to scroll the results. Larger tablets require longer, more sweeping gestures with which to execute the "pull."

! Caution

Although it's tempting to use it due to the pattern's sheer coolness, Pull to Refresh is not recommended for the majority of search results that deal with mostly static content. It is simply not satisfying to execute a pull and release and get the same data, and the watermark on the top of the list becomes chart-junk—a useless distraction. Other counter-indications of Pull to Refresh is for lists sorted in ways that do not lend themselves to rapidly updated content, such as Best Match, Price, and so on.

Here's another thing to keep in mind: The Pull to Refresh pattern is patented. That's right; Twitter currently holds the patent on this design. Although it's unlikely that Twitter would go after anyone other than a direct competitor using this pattern, it's an important caveat to keep in mind if you plan to use it in your app.

Related Patterns

None

7.5 Pattern: Search from Menu

Search is an option that can be accessed from the navigation bar menu.

How It Works

To do the search, the user must tap the menu button in the phone's navigation bar (that also houses the Back, Home, and Recents buttons) and then select the Search option. After Search has been tapped, the resulting page may show one or more of the following: saved searches, search refinement options, popular searches, nearby locations, and so on.

Example

In the Amazon app (see Figure 7.8), the customer accesses the search feature by tapping the magnifying glass in the menu located in the navigation bar.

FIGURE 7.8: The Amazon app uses the Search from Menu pattern.

The resulting Search page shows the previous query and a list of alternative query entry mechanisms, in this case a picture or a barcode that the customer can scan with an on-board camera. The menu is opened from the phone's navigation bar, which has been dynamically modified to add the app menu function.

When and Where to Use It

Despite being used by some of today's leading apps, this pattern is now largely deprecated. Most of the native Google apps in Android 4.0 have a dedicated Search button on the app's action bar or in the overflow menu (see the "7.6 Pattern: Search from Action Bar" section later in this chapter). Search from Menu is a transitional pattern that can still be used for a short time (or at least until the Android 4.0 Police show up) as a way to bridge apps in older Android versions with those in Android 4.0.

Why Use It

This is a popular pattern descended from older Android OS implementations, which recommended that the app's menu button always be present in the device's navigation bar. This handy pattern enables the designers to hide the search along with most of the rest of the navigation, on the navigation bar, which often eliminates the need for an additional action bar. This provides the advantage of a simple interface and "taller" vertical space so that more screen space is devoted to products or content.

Other Uses

Some older Android implementations, most notably those on the Motorola and LG hardware, provide a special dedicated hardware accelerator button for search. Tapping this button is the equivalent of tapping the menu button in the navigation bar and selecting Search from that menu.

This dedicated Search button has been removed from the latest hardware designed to run Android 4.0. You can speculate as to what this means long term, but in the immediate Android future, Search from Menu and Search from Action Bar search design patterns appear to take precedence over the dedicated hardware button.

Pet Shop Application

In implementing this pattern with the Pet Shop app, there are two options of what to put on the Search page. One option is to provide alternative input methods (refer to the Amazon app shown in Figure 7.8). Other popular options include previous searches and search refinements, such as filtering or sorting. Figure 7.9 shows previous searches.

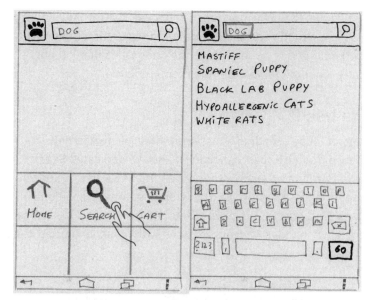

FIGURE 7.9: See the Search from Menu with previous searches in the Pet Shop app.

When showing the alternative query entry mechanisms such as barcode scan, picture, voice, NFC, and so on, recent previous searches can be shown as a grouped button (Recent Searches); although, this is generally less effective than actually listing previous queries in the list. Whatever strategy you decide to use, be sure to highlight (select) the current query as shown or provide an X or Clear button for the searcher so that starting a new search is easy.

Tablet Apps

Tablets do not generally need to use this pattern because there is plenty of room to install a dedicated search box or use the Search from Action Bar pattern instead.

Also, the Search from Menu pattern is ergonomically inferior to most other tablet patterns of search implementation because the menu button moves around constantly. In portrait mode the tablet's navigation bar is on the bottom of the device, which makes it generally awkward to access a menu from a normal tablet viewing position. (Read more about ergonomics in Chapter 3, "Android Fragmentation".)

! Caution

In addition to this pattern being deprecated in Android 4.0, using Search from Menu can lead to an awkward separation of the keyword query from the refinement tools. See the "7.9 Antipattern: Separate Search and Refinement" section.

Related Patterns

7.6 Pattern: Search from Action Bar

8.4 Pattern: Parallel Architecture

7.9 Antipattern: Separate Search and Refinement

7.6 Pattern: Search from Action Bar

The customer can access search via a dedicated button on the app's action bar.

How It Works

The Search button (usually styled as a standard Android magnifying glass icon) is shown on the top or bottom action bar. After the user taps Search, the resulting page shows one or more of the following: saved searches, search refinement options, popular searches, nearby locations, and so on.

Example

Google Plus offers an excellent example of this pattern (see Figure 7.10).

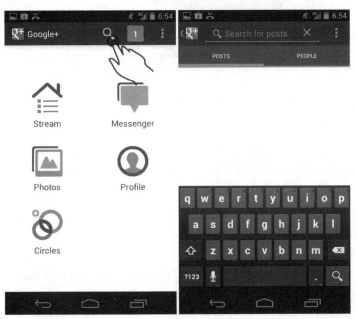

FIGURE 7.10: The Google Plus app uses the Search from Action Bar pattern at the top of the app.

Google Plus offers a dedicated Search button on the top action bar. Tapping the Search button navigates the user to the dedicated tabbed search page, with two search subdomains, Posts and People, displayed as tabs. Tabs are a common pattern in search, as discussed in Chapter 9, "Avoiding Missing or Undesirable Results."

Another example of the dedicated Search button in the action bar is in the Android Messaging app.

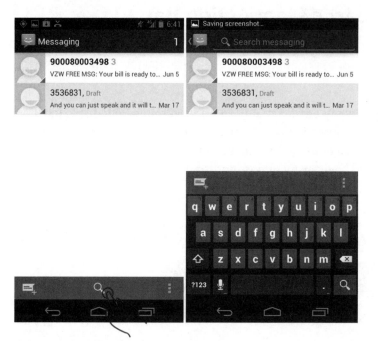

FIGURE 7.11: The Android Messaging app includes the Search from Action Bar pattern in the split action bar at the bottom of the screen.

In the Messaging app, the Search button is in the middle of the split action bar, which is at the bottom of the screen. Inconsistent? Sure. But relative freedom of placement of controls on the screen is a large part of the Android DNA (refer to Chapter 2, "What Makes Android Different").

When and Where to Use It

Any time you have an action bar in your app that has some space on it and search is important to your customers, this pattern is a great choice. Ergonomically, placing the Search button on the bottom of the split action bar makes it easier to access the function one-handed.

Why Use It

Although I am not aware of any official standing on the matter, it seems that the Google Android team has made a real effort to generally replace the Search from Menu pattern with the Search from Action Bar pattern, at least in native Google apps in Android 4.0. This is a strong signal that search remains important at Google. If search is likewise important to you, this pattern is an excellent choice and is now more or less "official" (to the extent that anything in Android can be considered official).

Other Uses

When the app's screen real estate shrinks due to the size of hardware that runs it, some action bar functions may move into the overflow menu, as discussed in Chapter 1, "Design for Android: A Case Study." In this case, the search function shown on the action bar might be forced into an overflow menu as well. To access the search function, the customer will have to tap the overflow menu and select Search—pretty straightforward.

Pet Shop Application

Contrast the Search from Action Bar pattern shown in Figure 7.12 with the Search From Menu pattern referred to in Figure 7.9.

FIGURE 7.12: Check out the Search from Action Bar in the Pet Shop app.

Both patterns enable access to the search page from anywhere in the application and use the same search page design. However, with the Search from Action Bar pattern, getting to the search page is accomplished via a single tap on the dedicated Search button on the App bar rather than in the two taps required by the Search from Menu pattern. Search from Action Bar saves an extra tap and surfaces the search much more prominently in the mind of the customer. There is a drawback, however; using this pattern adds an action bar, which takes away precious pixels from the vertical space available for viewing content and products.

Tablet Apps

This is the standard search pattern to use in tablet apps. However, if you use the standard top action bar layout that places the search icon somewhere close to the middle of the action bar (refer to the Messaging app in Figure 7.11), your customers may get a severe case of what Josh Clark has dubbed "Tablet Elbow" if they must tap this button often (read more in Chapter 3). A better placement of this button is on the right or left nav bars, which run vertically along the edges of the device (see Chapter 14 to find out more about tablet-specific patterns).

! Caution

Similar to Search from Menu, Search from Action Bar can also lead to an awkward separation of the keyword query from the refinement tools. See the "7.9 Antipattern: Separate Search and Refinement" section.

Related Patterns

7.5 Pattern: Search from Menu

8.5 Pattern: Tabs

7.9 Antipattern: Separate Search and Refinement

7.7 Pattern: Dedicated Search

The search box is placed on top of the search results and does not scroll with them.

How It Works

The search box sits on top of the search results, which enables customers to easily edit and fine-tune the keyword query. Often, a refinement (filter) button is placed to the left or right of the search box.

Example

A great example of this pattern is Yelp, as shown in Figure 7.13.

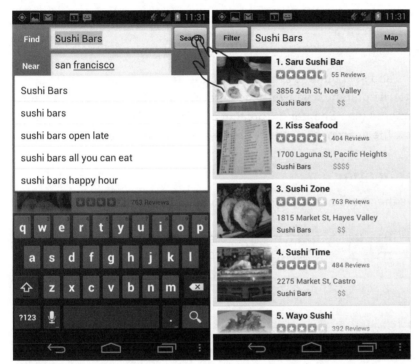

FIGURE 7.13: The Yelp app includes a good example of the Dedicated Search pattern.

The dedicated search box in Yelp sits on top of the search results and does not scroll when the search results are scrolled. In addition, search tools, such as Filter and Map, are located on the same line as the search box.

When and Where to Use It

For apps in which search is a key part of functionality, the Dedicated Search pattern is an excellent choice. The Dedicated Search pattern shows clearly what keyword query yielded the search results and provides convenient, dedicated tools to change the query and access other refinements.

Why Use It

As Peter Morville and Jeff Callender so eloquently stated in their book *Search Patterns* (O'Reilly, 2010), "What we find changes what we seek." Nowhere is this statement truer than in the mobile space, where typing is awkward and people are highly distracted with multitasking. People prefer to start general and refine rapidly, and changes to the keyword query are part of that refinement. The Dedicated Search pattern addresses the need with unmatched simplicity and elegance. The original keywords that the searcher types are always visible on top of the results and are retained in the search box for easy editing.

Other Uses

If additional filters and sort options are used with the keyword query, the Dedicated Search pattern combines well with the Filter Strip pattern that shows filters and query refinements (see Chapter 8, "Sorting and Filtering"). Together, these two patterns show the searcher the entire contents of a complex query.

Pet Shop Application

Figure 7.14 shows the implementation of the Dedicated Search pattern.

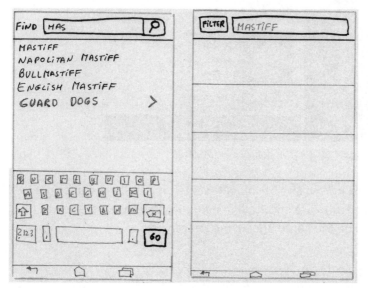

FIGURE 7.14: This is how the Dedicated Search pattern looks when used in the Pet Shop app.

This is a fantastic pattern for the Pet Shop app if you expect customers to edit their queries often.

Tablet Apps

Tablets are much less screen space–challenged than mobile phones. For most apps that use search, having a dedicated search box is an excellent idea. Simply having a dedicated search box on top of every page in the app implements the Dedicated Search pattern nicely.

! Caution

Having a dedicated search box on top of the page does not mean that you need to give up the person's history of previous searches or auto-correct functionality. Remember that previous searches can be easily presented via a layer under the search box (refer to the "7.2 Pattern: Auto-Complete and Auto-Suggest").

On smaller devices this pattern takes up a fair bit of vertical space (20 to 30 percent of the total screen space), which significantly reduces the number of products or the amount of content that can be shown to the customer. The Dedicated Search pattern is akin to reducing the number of books that can be shown on a bookstore shelf because of the giant sign that tells you the name of the section. It's not always a bad thing, but it is something to keep firmly in mind.

Related Patterns

8.3 Pattern: Filter Strip

7.2 Pattern: Auto-Complete and Auto-Suggest

7.8 Pattern: Search in the Content Page

The search box is on top of the search results and part of the content page, so it scrolls with the rest of the content. This pattern is an alternative of the Dedicated Search pattern.

How It Works

The basic premise of this pattern is that the search box is part of the content page. When the page first loads, the search box is shown to the customer. As the customer scrolls the content page down, the search box simply scrolls out of view with the rest of the content. To search, the customer must scroll back to the top of the page.

Example

The Twitter app makes an effort to have a consistent interface on iOS and Android, which makes it a good example of the Search in the Content Page pattern (see Figure 7.15).

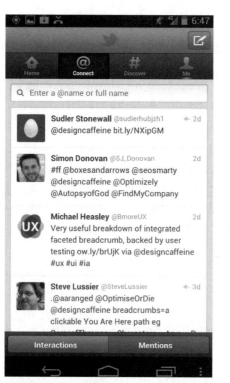

FIGURE 7.15: This is how the Twitter app uses the Search in the Content Page pattern.

This pattern works well with the Pull to Refresh pattern described earlier.

When and Where to Use It

Any time you have a screen that is content-centric but might need to be occasionally searched, Search in the Content Page is a great option. However, make sure that your customers want to only run keyword queries and that sort order is obvious and does not need to be changed. Ideally, people never want to have any refinement on the query because this pattern generally makes search refinement awkward.

Why Use It

This pattern is popular in iOS but is currently seldom used in Android. That's a shame because it's ideal for certain applications. In particular, content-centric screens such as name lists or activity streams such as updates, which are normally browsed but not searched, make great candidates for use of this pattern.

The Search in the Content Page pattern makes search easily available but does not take up permanent screen space the way the Dedicated Search pattern does.

Other Uses

One modification popular in iOS but virtually unknown in Android is Scroll to Search. When a content page loads, a search box is hidden on top of the page. Pulling the page down reveals the search box that searches within the content on the page. After the query runs, the resulting page shows the search box with the query.

Pet Shop Application

This pattern is not suitable for e-commerce because it makes refinement awkward. However, you can use it for an update stream or Pet News section, where search is likely to be infrequent and made up of keyword queries (see Figure 7.16).

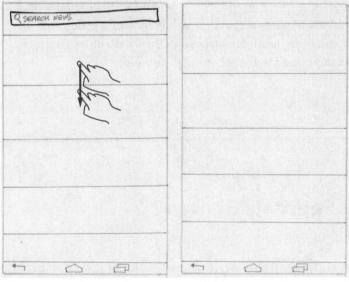

FIGURE 7.16: The Search in the Content Page pattern appears in the Pet News section of the Pet Shop app.

Tablet Apps

This pattern is all about saving space, which makes it superfluous for tablets, which generally have enough space. However, it still has its place because it is easy to implement.

! Caution

This pattern is currently rare on the Android platform but is quite widespread on iOS. The reasons for this are not clear. One possibility is that iOS enables a quick scroll to the top of the page (which thereby "jumps" to the search box) using a single tap in the middle of the top App bar. This single tap jump to the top of the page shortcut is unavailable on Android because the top of the screen is normally occupied in the Android OS by the Notifications strip, which understands the pull-down touch gesture. This could make frequent use of search functionality in Search in the Content Page implementations problematic on Android because the person must deliberately scroll back to the top of the page "the long way" to reveal the search box.

For the *Scroll to Search* modification of this pattern described in the "Other Uses" section, the reason could be even simpler but more insidious. Although at this time I'm not aware of any limitation, Apple could be holding a patent to this pattern, so Android apps are generally prevented from using it (or it could be more popular in iOS simply from lack of screen space, which is less of a problem with larger Android devices). If you're in doubt, use the simple version of this pattern implemented by Twitter as described in the "Example" section.

Related Patterns

7.7 Pattern: Dedicated Search

7.4 Pattern: Pull to Refresh

⊖ 7.9 Antipattern: Separate Search and Refinement

An awkward experience results when the keyword query search box is removed by two or more taps from the other search refinements.

When and Where It Shows Up

Any time the keyword query and multiple complex refinement options are separated, you must pay attention. Although this shows up frequently on iOS, this antipattern is especially an issue on Android because of the widespread use of dedicated search pages, the result of Search from Menu and Search from Action Bar patterns.

Example

It's easy to mess up when blindly copying successful apps and applying a slightly different paradigm. For example, the Amazon app manages to pull off using Search from Menu and a separate keyword search page successfully by using a simple filter drop-down located in-page with the rest of the content (refer to Figure 7.8).

Contrast the Amazon app search and filter scheme with that in TheFind, as shown in Figure 7.17.

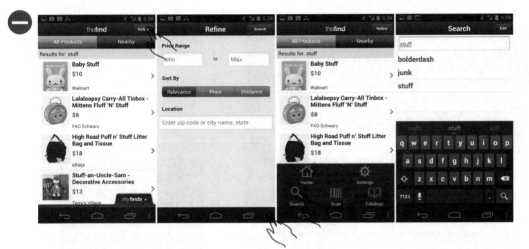

FIGURE 7.17: There's an awkward separation of keyword search from the rest of the search refinements in this antipattern from TheFind.

The refinement page is a dedicated page with multiple text fields. One thing is conspicuously absent: the keyword search box. To change the keywords in the query, the user must tap the Menu button and then tap Search. This separation is completely artificial and therefore awkward, which should be avoided.

Why Avoid It

In most people's minds, search is an iterative activity. (Recall Peter Morville's quote, "What we find changes what we seek.") So in the mind of searchers, there is little separation between keywords, filters, and sort options. These are all tools to find what they want. Separate Search and Refinement is an antipattern precisely because it introduces awkward separation between the keyword query and everything else. This is neither wanted nor needed. Separate Search and Refinement breaks the association between different parts of the query and makes it difficult to find what you want and stay in the flow.

A better pattern called Parallel Architecture or any of the simple faceted search patterns covered in Chapter 8 offer a more usable configuration.

Additional Considerations

Although often harder to recognize, the Separate Search and Result antipattern also occurs when search is presented in a different way on the homepage and on a separate search page. For example, TheFind app also offers a different search from the homepage shown in Figure 7.18.

FIGURE 7.18: In this antipattern there are two slightly different places for a keyword search in TheFind homepage and dedicated search page.

Although it has similar search functionality at first glance (neither one have any refinements, for example), the homepage search doesn't have the previous search's history widget that the dedicated Search page has. This "separate homepage search" antipattern is a child of the Separate Search and Refinement antipattern. It can be daunting to customers who quickly get lost.

Unfortunately, this situation happens quite often and is much harder to recognize and prevent. Two great solutions for this issue are the Parallel Architecture pattern, where the homepage *is* the basic search page, and the Dedicated Search pattern, which presents a consistent search box and functionality on the homepage and search results pages.

In general, it's a good idea to offer the same basic search functionality every time you have the search box. If you offer history and auto-suggest in one place, do it everywhere you use the basic search box. Also, avoid having multiple places for search that differ only slightly; it makes it too easy for people to get lost and confused and abandon search altogether.

Related Patterns

8.4 Pattern: Parallel Architecture

7.5 Pattern: Search from the Menu

7.6 Pattern: Search from Action Bar

7.7 Pattern: Dedicated Search

CHAPTER 8

Sorting and Filtering

After your customers get the information scent that lets them know they are on the right track to finding what they need, the next step is to help them winnow down the avalanche of data to the size manageable in the mobile context of use. That's where the sorting and filtering design patterns in this chapter prove invaluable. Many of the so-called Experimental Patterns, such as Sliders with Histogram (Chapter 10, "Data Entry") and Filter Accelerators (Chapter 11, "Forms") can also be used for sorting and filtering, so be sure to also check out relevant sections in those chapters.

⊖ 8.1 Antipattern: Crippled Refinement

Before talking about refinement patterns, let me make something very clear: Mobile web should be as good as desktop web, or better. Crippled Refinement is a fundamental UX antipattern to be avoided at all costs.

When and Where It Shows Up

Whenever there is an option to filter and sort mobile or tablet search results, there is a great temptation to "dumb down" the interface to only one or two refinement options. Additionally, mobile interfaces that limit search refinement to a single step fundamentally cripple the flow of search experience, which proceeds via multiple consecutive refinement steps.

Example

Although it is overall a highly successful and lucrative mobile app, Amazon provides the example of a dumbed-down filtering process.

Just compare the *only* refinement option, Department (see Figure 8.1), available in the mobile app to the multitude of filters and sort options available for the same "Harry Potter" query in a desktop browser, as shown in Figure 8.2.

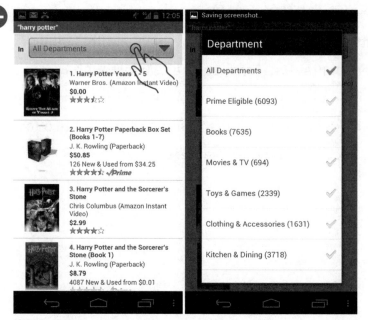

FIGURE 8.1: The Amazon app includes an example of the Crippled Refinement antipattern.

FIGURE 8.2: Here's a battery of filters and sort options in the Amazon.com desktop web.

The mobile app seems rather seriously deficient, doesn't it? But the real, more insidious UX problem is deeper still. Typical search flow involves multiple steps of refinement and changes, with each step directly informing what the customer will do next. Because the Amazon mobile app refinement is designed to be only a single-step operation, the interface does not support any changes to the original keyword query. Actually, any changes to the query *reset* the entire search (see Figure 8.3).

Referring to the sequence of steps pictured in Figure 8.3, the searcher started the search broadly, by searching for Harry Potter. Remembering that his niece wanted a DVD, he picks a category Movies & TV. Viewing the items in the search results, the searcher realizes that he wanted to see all the buying options for the *first* movie, *Harry Potter and the Sorcerer's Stone*, so he taps the Menu button and selects Search once more, updating the keyword query to include the full name, Harry Potter and the Sorcerer's Stone. At this point, the searcher unexpectedly finds his category reset back to All Departments—his refinements gone, his discovery flow interrupted.

The best mobile interfaces support multistep refinements, allowing the customers to do more, not less, than the desktop interface.

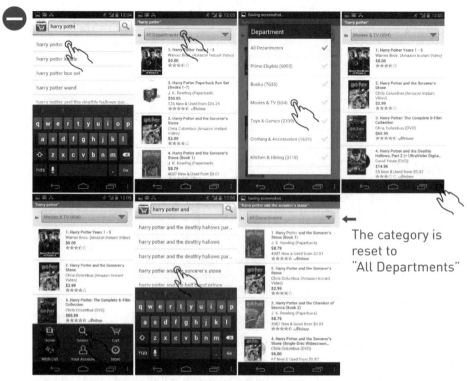

The category is
reset to
"All Departments"

FIGURE 8.3: The Amazon mobile app does not support multistep refinement.

Why Avoid It

Edward Tufte (the "Da Vinci of Data") famously quipped, "Clarity and simplicity
are completely opposite of simple-mindedness." Nowhere is this adage more true
than in mobile and tablet design. Mobile and tablet apps must support most if not
all the desktop web filters and make available what you know to be a multistep
refinement finding flow.

If you decide to provide less functionality simply because you think it is too much
to handle for mobile customers, think again. Experience with crippled apps such
as the early version of Facebook clearly shows that people want to do *more* with
mobile apps, not less. Instead of dumbing down, think of the way to provide filtering and sorting functionality that works with the on-board sensors and device
limitations.

The key to great mobile design is not really having customers *doing* more (that is
more work), but *getting* more: more capabilities, more value out of the experience.
Well-designed mobile interfaces actually enable people to do *less* work because
you can leverage other sensor data to infer more of the customer's context. You
could say people want more or expect more from their mobile device, which means

they want to *do* less. The key is context and real human desires, not doing more or less because of some abstract principle. Use the patterns in the rest of this chapter as a guide and be sure to test early and often with your target customers.

Related Patterns

All patterns in this chapter

8.2 Pattern: Refinement Page

Search results enable access to sort and filter options on a separate page or *lightbox*.

How It Works

When the customer wants to refine the query, he can access a dedicated page or a lightbox with filter and sort options, followed by some sort of a "go" button to re-initiate the now refined search. Optionally, the set of refinements applied to the search might be shown to the customer with a Filter Strip design pattern. (Filter Strip is covered next, in section 8.3.)

Example

One excellent example of a full-featured mobile refinement is the eBay app. (See Figure 8.4.)

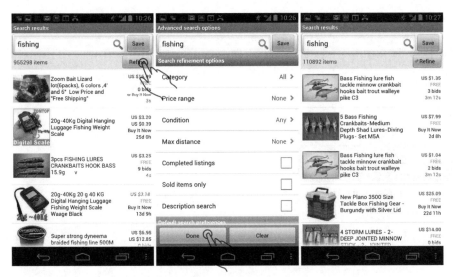

FIGURE 8.4: The eBay app is an excellent example of a Refinement Page pattern.

Despite apparent complexity, this remains one of the most useful and profitable mobile apps, with several *billion* mobile e-commerce dollars generated as of the date of this writing. One reason for its success is that designers refused to dumb down the experience and have instead done an excellent job managing complexity: The app provides a full-featured eBay search experience in a tiny mobile device.

This app makes an excellent example because it shows off some uncommon but useful refinements, including

- Multiple select check boxes (Free Shipping, US-only) as shown in Figure 8.5

- Multistep drill-down refinement (Refine › Category › Sporting Goods) with additional subcategories (see Figure 8.6)

- Sort on the same refinement page as search (Best Match and so on) as shown in Figure 8.7

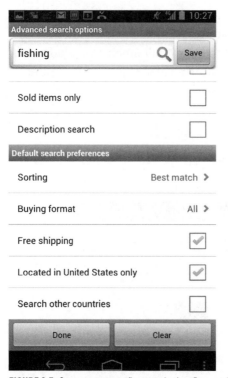

FIGURE 8.5: One uncommon refinement in the eBay app is the use of multiple select check boxes.

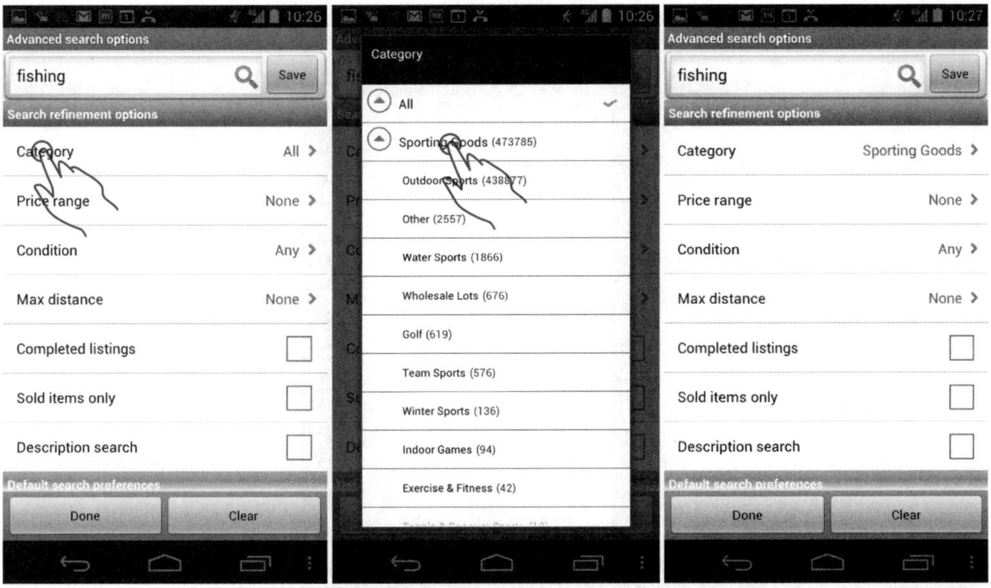

FIGURE 8.6: This is an example of drill-down refinement.

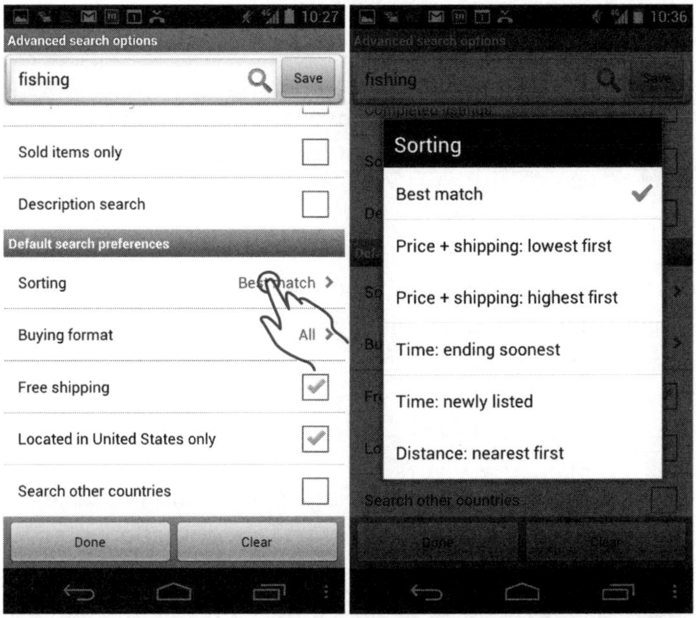

FIGURE 8.7: The eBay app provides both search refinement and sort in the same Refine page.

But the eBay app not only offers full-featured refinements, but also actually does *more* than the desktop web version—for example, offering refinement by distance from the current location without having to enter the ZIP code (see Figure 8.8).

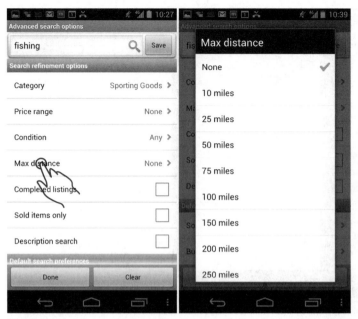

FIGURE 8.8: The Refinement Page pattern in the eBay app offers distance-based refinement.

When and Where to Use It

Any time you have a faceted search or there is a use case for changing the sort order, the Refinement Page pattern is an excellent tool for the job.

Why Use It

Refinement is an essential part of search, especially on the mobile device where typing avoidance generally gives rise to shorter, more general queries, which need to be refined quickly and efficiently.

Other Uses

Some of the best implementations of the Refinement Page pattern are not pages at all. They are Refinement Page lightboxes (also somewhat loosely called pop-ups). Early Android phones simply did not have sufficient real estate to implement refinement in a lightbox that could still be operated with fat fingers, and the iPhone's smaller size still prevents iOS designers from doing this effectively in most cases. However, today's higher-end Android devices have a decent-size screen, which means multiple refinement options can be displayed in a lightbox, as shown in Figure 8.9, demonstrating the excellent Yelp refinement example.

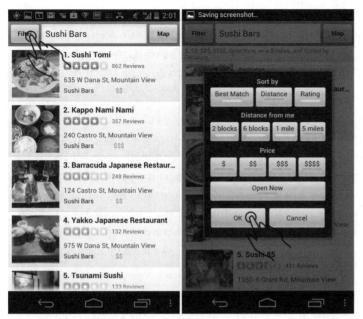

FIGURE 8.9: The Refinement Page Lightbox in the Yelp app is a good implementation of the pattern.

Why is there a distinction between lightboxes and dedicated pages? The reason has to do with maintaining flow within the current task. As discussed in greater detail in Chapter 13, "Navigation," immersive UIs (of which Refinement Page Lightbox is a good example) *maintain the illusion of staying on the same page and within the same task*, whereas separate pages appear to *switch* the task from one interacting with the content to that of refinement. Separate pages yank the person out of the search results and place her in a different environment. If you can accomplish your refinement tasks using a lightbox without sacrificing functionality, it almost always creates a better experience.

Pet Shop Application

Either the dedicated refinement page or lightbox refinement could be usable for the faceted refinement and sorting in the Pet Shop app. What makes the design of the refinement functionality even better is the addition of the Filter Strip pattern. In the next section, the entire package is wireframed together using the agile sticky notes methodology.

Tablet Apps

Refinements are important for tablets. Tablets have more surface area, which exposes more inventory to the customer and presumably causes them to make the decision to refine even sooner. Tablet owners also presumably care more about flow than the mobile users, and tablet search tasks last *on average* longer than similarly structured mobile tasks. Together these factors clearly dictate that the tablet refinements need to be as full-featured as web refinements, *plus* provide the refinements that are only available on touch devices (such as using the GPS locator feature for distance filtering).

Because tablets have more surface area, most refinements should be done via lightboxes, not dedicated refinement pages. Remember to place the lightbox in the correct ergonomic position where the controls can be manipulated easily in both portrait and landscape orientations. In the absence of specific use information, the best place is typically on the top-right corner of the screen, where inputs in the light-box can be manipulated with the right thumb without the user having to let go of the device completely or while using a typical two-handed grip with the bottom of the tablet resting on the user's lap or a table (see Chapter 3, "Android Fragmentation").

! Caution

With the increase in screen size, there is a strong trend in many apps, such as Kayak, to have separate filtering and searching controls (large buttons on the bottom of the screen), as shown in Figure 8.10.

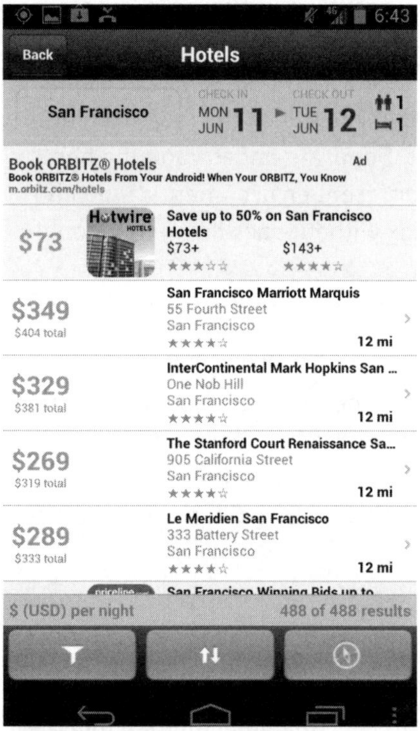

FIGURE 8.10: The Kayak app includes separate Sort and Filter refinement controls.

This is a mistake. As described in *Designing Search: UX Strategies for Ecommerce Success* (Wiley, 2011, ISBN: 978-0-470-94223-9), my original research shows that most people have trouble differentiating sorting from filtering. Furthermore, this distinction is not nearly as important as most designers and engineers think. Consider that most people will never see more than the first 100 to 300 items of a typical result set, especially on a mobile device. In other words, people can never hope to view anything but a tiny fraction of today's typical high-volume result set numbering in the thousands. The result is *The Mystery of Filtering by Sorting*. In many people's minds they *filter* out high-price items when they *sort* by Price: Lowest First, because they can never go through the entire result set to view the high-price items at the bottom of the list. Thus the action of sorting creates an inherent filtering effect due the large number of items in the set.

On mobile devices with limited screen real estate, filtering and sorting options are often placed on the same refinement page or lightbox. (Both the Yelp and eBay apps do this.) This is a great trend that should be encouraged because it matches well with the user's mental model of these controls, which are both seen by customers as types of "refinement." Furthermore, as I explain in my first book, sorting controls should often be shown first and called out especially because sorting often creates a much better refinement of the result set: It never causes zero results; its outcome is predictable; and it's hard to screw it up. So, in short, you should consider combining filtering and sorting under a single Refine Area, unless there is a good reason to do otherwise, and make refinement as much of an integral part of finding, in general, as possible

Last but not least, faceted filtering cannot strictly be considered being faceted without letting the customer know the numbers of items that will result when the facets or filters are applied. Numbers of items associated with refinements let customers know how to proceed. Unfortunately, knowing numbers of items is particularly hard to do with multiple refinements being applied in a single shot using check boxes, as they are in the eBay and Yelp apps. By default, this creates a refinement experience that is inferior to that of using a desktop web browser. In your apps, if at all possible, try to indicate the outcomes of user refinement actions, including how many items will be left over after the filter is applied in order to avoid zero-results outcomes.

Related Patterns

8.3 Pattern: Filter Strip

13.5 Pattern: Watermark

8.4 Pattern: Parallel Architecture

8.3 Pattern: Filter Strip

A small horizontal section of the screen shows the keyword query and the filters applied.

How It Works

After the search is run, the entire contents of the query are shown to the searcher in a thin Filter Strip. Any changes to the query (such as filters applied, sort order, and so on) are continuously reflected in the Filter Strip that acts as an informational display to show the searcher the details of the exact query he applied to his search.

Example

Yelp is a good example of this pattern. Because the Yelp app is an example of the Dedicated Search pattern (read more in Chapter 7, "Search"), showing the keyword query again is not necessary. Instead, Filter Strip (see Figure 8.11) shows up only when additional refinements (filters and sort order) are applied to the search.

Note that the Filter Strip text is small and that it wraps around in the rare cases in which this is warranted by the number of refinements applied.

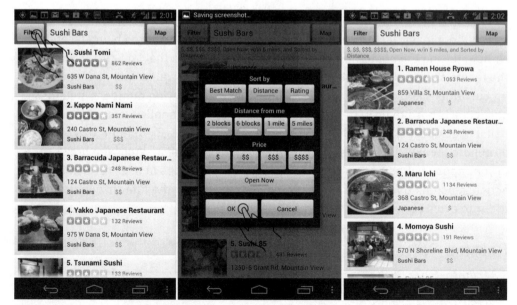

FIGURE 8.11: The Yelp app uses the Filter Strip pattern when refinements are made to a search.

When and Where to Use It

Any time a set of filters or sort refinements can be applied to a query, consideration of the Filter Strip pattern is warranted, regardless of whether the search box is a dedicated one (as it is in Yelp) or if search is initiated in some different manner as discussed in Chapter 7.

Why Use It

Knowing "where you are" is a continuous challenge in mobile space with its small screen and constant interruptions. Having a solid reliable way to communicate all the search parameters to the searcher enables him to adjust search parameters, sort order, and refinements until the query results are satisfactory. The Filter Strip pattern enables designers to effectively accomplish this in an unobtrusive way that takes up only a fraction of the screen real estate.

Other Uses

Filter Strip in its *best* implementation is semi-transparent, as shown in the iOS Yelp app in Figure 8.12.

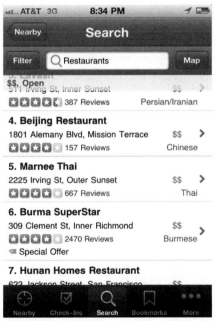

FIGURE 8.12: In the Yelp iOS app, the Filter Strip is semi-transparent.

In contrast, the majority of Android implementations of the Filter Strip are solid. This is a mistake. Making the Filter Strip semi-transparent enables the searcher to read the search results while *also* seeing the query clearly, especially while scrolling the page. If the Filter Strip is semi-transparent, the focus shifts from the process of search and refining to what is most important to the searcher: search results content. Furthermore, the screen space is used efficiently—not a pixel is wasted. This is important for immersion as discussed in more detail in Chapter 13. For now, suffice it to say that successful games such as Angry Birds use semi-transparent controls frequently, and you should consider the lessons from the games to improve the effectiveness of the search experience in your own app, while balancing immersion with competing UX concern of legibility and perceptual noise.

Pet Shop Application

Filter Strip is perfect for the Pet Shop search because of the many refinements that can be applied. The example of the complete search refinement and filter strip package (first introduced in the discussion of the Refinement Page pattern earlier in this chapter) is wireframed, as shown in Figure 8.13, including a semi-transparent Filter Strip that appears after the refinements have been applied.

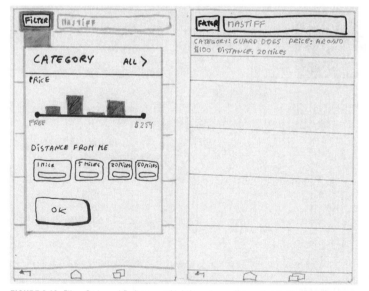

FIGURE 8.13: Filter Strip and Refinement Lightbox play well together in the Pet Shop app.

Several unusual refinements are used, such as a dual slider with histogram to filter by price. To learn more about this Slider experimental design pattern, head over to section 10.1. Also note that the lightbox comes out of and is connected to the Filter button—this is deliberate because it projects a strong perception of the app's Information Architecture (IA) through intuitive visual means. Maintaining this connection clearly states, "This lightbox is the contents of the Filter expanded." Having the lightbox connected with the element from which it originated also dispenses with the need for a dual Go and Cancel button arrangement in the lightbox, as shown in Yelp. Instead, to exit the lightbox, the searcher needs to tap only the Filter button again to collapse the lightbox. This design greatly decreases any accidental fat-finger taps on the Cancel button and increases "fool-proofness" and satisfaction. Use the custom Expand and Collapse animation effect (where the lightbox literally "grows out" of the Filter button) for the open lightbox transition to further support the sense of place and provide added visual slickness.

Last, but not least, note that the Filter Strip reports that the price has been applied as "around $100"; this is deliberate. The additional precision that is so important for computers carries little meaning for human beings, so it pays to shorten the numeric filter settings to human-scale formats such as "around $100," especially on mobile devices and tablets. For more on this, see Chapter 8 of my book *Designing Search: UX Strategies for Ecommerce Success* (Wiley, 2011, ISBN: 978-0-470-94223-9)

Tablet Apps

Filter strip is just as important for tablets. You do not need to use as small a font as is typical for the mobile devices because tablets have more space and are meant to be read comfortably at a longer distance from the eye. Transparency of the Filter Strip element is also less of an issue; although it's still recommended, especially if the search results are more visual in nature, such as large pictures. You want as little as possible to detract from the content, and semi-transparency of the filter strip helps it "disappear," which encourages immersive flow interaction with the content.

! Caution

Use text that is as small as possible, but don't lose legibility. To do this effectively, avoid temptation to use the Filter Strip as a clickable object or button, unless it is at the very top or very bottom of the screen. The main purpose of the Filter Strip in most cases is to provide information while remaining unobtrusive; it's not to be clicked.

Related Patterns

7.7 Pattern: Dedicated Search

13.6 Pattern: Swiss-Army-Knife Navigation

13.5 Pattern: Watermark

8.4 Pattern: Parallel Architecture

"Basic" and "advanced" search screens form two parallel tracks to search results.

How It Works

Results can be accessed in two ways: via simple search and via advanced "more options" search. Tapping the back button navigates to the original search starting screen and enables searchers to switch mode. There may also be a separate "refine" button on the search results screen that enables the searcher to enter the advanced mode.

Example

Yelp forms an excellent example of this pattern. If the searcher is simply looking for a place to grab a cup of coffee, she can easily tap "Coffee" on the home screen and see the nearest places to get caffeinated. However, if the searcher is adamant that she wants the best, she can take the time to type in "Peet's Coffee" into the search bar, mixing a few more constraints into her coffee cup (see Figure 8.14).

When and Where to Use It

Use the Parallel Architecture pattern whenever search conditions might be served better through minimal search input or browsing, while at the same time there exists a common use case for more advanced, full-featured search.

Why Use It

The classic mobile use case for this pattern is *local*. Whenever there is a strong local component, consider using this powerful and intuitive "mobile-first" pattern. Using this pattern requires little mastery of the app or subject matter and makes it easy to engage with your content right away, while providing people who want something specific with a powerful set of tools to refine and customize their queries.

Other Uses

Local is not the only use case for basic versus advanced search. You can argue that apps such as Facebook and Twitter also follow this pattern with the recent updates you can browse without typing, while also providing a feature to search the updates. However, one important requirement for this pattern is to present both basic and advanced search results in mostly the same way (maybe with additional keyword highlighting for the keyword search versus browse). Otherwise, you have a separate search functionality.

FIGURE 8.14: The Yelp app is an excellent example of Parallel Architecture.

Pet Shop Application

Simple browsing is not the only way to start basic search. Using the ThirstyPocket Labs iPhone app as a model, you can see how easy it would be to provide multiple levels of engagement in the Pet Shop app. The wireframes in Figure 8.15 provide two screens: Home and Advanced Search, which I took the liberty of renaming to More Search Options because my research showed that label to be less intimidating.

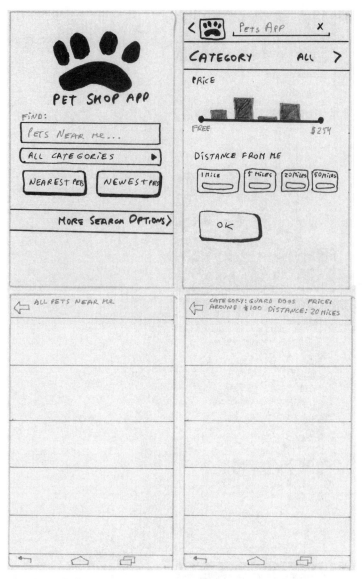

FIGURE 8.15: The Parallel Architecture pattern provides multiple convenient engagement levels in the Pet Shop app.

From the home screen, the searcher has several simple engagement options:

- Browse pets for sale in her neighborhood by tapping the Nearest Pets button (sorted by nearest first).

- Browse the newest pets on the market in a reasonable driving radius of 50 miles using the Newest Pets button (sorted by newest first).

- Browse pets by category (sorted by nearest first or newest first, depending on which search button is tapped).

- Combine category and keyword search (sorted by nearest first or newest first, depending on which button is tapped).

And by tapping More Search Options, the customer can engage with a dedicated page that has a variety of additional filters such as price and distance, for those customers that use their mobile device to search for pets from a different location than their home (while they are at work, for example) or have special needs and budget constraints not covered by the "basic" home search screen.

Another thing to note in the design in Figure 8.15 is the addition of a tappable semi-transparent action bar above the search results. The action bar includes Filter Strip reporting functionality, yet also provides a single-function back button. Because this action bar is semi-transparent, it does not take any screen space away from the search results, which occupy 100% of the screen. More discussion on screen real estate use in games and other immersive experiences is in Chapter 13.

Tablet Apps

Although local is not as big a deal for tablets as it is on smaller mobile devices, simple "browse" engagement that enables people to dive straight into the featured content without having to enter anything is arguably even more important for tablet than it is on mobile. This means that the Parallel Architecture pattern is an excellent option for tablets. Consider just one simple example of reading News: In the "basic" view the person can explore the latest news or news by topic. In the "advanced" view, she can search by keyword, category, publication, and date range to drill down further into a particular topic of interest.

! Caution

Although the Parallel Architecture pattern appears to be simple and versatile, apparently there exists nearly endless opportunities to mess it up. One popular way to use Parallel Architecture is to add another Search from the Menu search box to the application that already has a basic and advanced search, creating a completely different UI for viewing and refining the same data. One unfortunate victim of this is the TripAdvisor app, which appears to be confused about the goals of its customers and its IA.

As you can see from Figure 8.16, there appear to be multiple ways to access and refine essentially the same content, which makes a simple task of finding a local hotel a Herculean Labor comparable to cleaning the famous Augean Stables.

Problems begin on the home screen. Tapping a hotels icon takes the searcher to what would appear to be a "basic" search screen of the Parallel Architecture pattern. However, instead of making the local search a default (as they should almost always be in any mobile travel app) TripAdvisor has added the Local Search option as a small check box. The check box control is both hard to find and hard to tap with the finger.

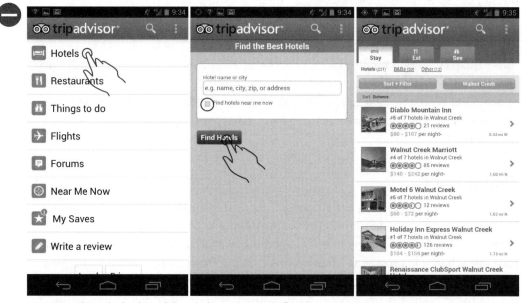

FIGURE 8.16: Accessing local search from the home screen in the TripAdvisor app is more complicated than it should be.

The second way to access the same content is with a Search from action bar search icon. (See Figure 8.17.) This search appears to be tacked on as an afterthought, making the results of even the simplest queries like "Hotel" initiated from the search box perfectly useless. The filter pop-up is not designed to be of any help, offering only a few meager and confusing options (which do not even appear to be working as of the date of this writing). Also unexpectedly missing on the resulting search results page are the three tabs on top of the other set of search results. Instead the tabs appear as options in the filter pop-up; although they are named differently (Restaurants" versus "Eat", "Things to do" versus "See", and so on) and sport icons visually distinct from the set seen in all other search results. The filter pop-up also adds the fourth option, "Locations," adding to the general confusion.

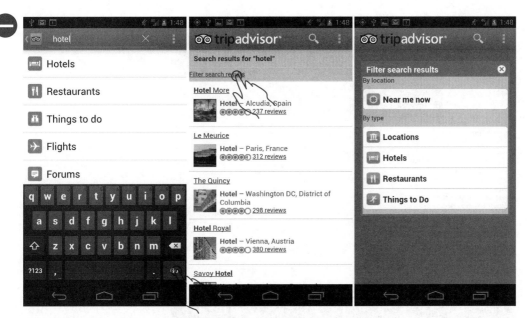

FIGURE 8.17: Accessing local search from the App bar in the TripAdvisor app results in a different interface and refinement options.

But that's not all. There exists yet another way to access content: via the menu's Near Me Now option. (See Figure 8.18.) This browse function at first glance holds the most promise, yet it is nearly impossible to find. When tapped, Near Me Now unexpectedly catapults the searcher into the second search results tab called "Eat" (regardless of the previous searches that were all for hotels.)

The designers of the TripAdvisor app probably didn't aim to create this gargantuan mess. The confusing IA may have come about as a result of responding to the criticism of "people can't find our stuff" via a knee-jerk reaction of adding multiple, and ultimately confusing, ways to find the same content.

In your own app design, use the patterns in this book as a guide, but realize you will never please everyone. When the inevitable criticism comes in, rather than adding more features, maintain a clear vision of supporting your primary use cases while continuously doing field research to figure out where your best customers are getting stuck. Knowing the facts on the ground will give you ammunition to defend your design decisions and confidence to stick to your vision.

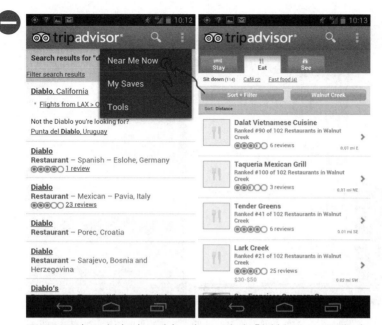

FIGURE 8.18: Accessing local search from the menu in the TripAdvisor app catapults the user to the "Eat" tab.

Related Patterns

8.3 Pattern: Filter Strip

8.2 Pattern: Refinement Page

7.5 Pattern: Search from Menu

8.5 Pattern: Tabs

Tabs on top of the page enable customers to switch views or apply popular sort and filter options.

How It Works

When the search results are shown to the customer, the results are further seg-regated into two or more views shown as tabs on top of the page. The first tab is usually the default result. Tapping one of the other tabs switches to a different view of the data. For example, tabs can be used to toggle between list, gallery, and map views of the same set of results.

Example

Wikitude is an app that uses the tabs design pattern to switch between views. Wikitude recently redesigned the app for Android 4.0 OS (see Figure 8.19).

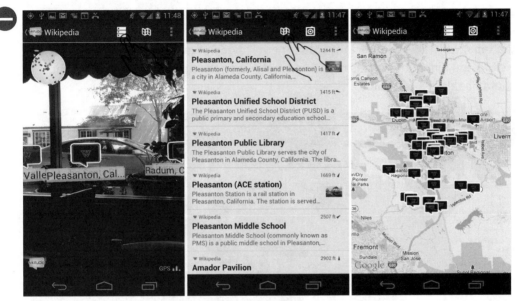

FIGURE 8.19: The Tabs pattern is used in the Android 4.0 Wikitude app.

Tabs are shown as miniature icons. The tabs look considerably more sophisticated and visually polished than the older version (shown in Figure 8.20) and take up little screen space. The default setting displays the search results in an Artificial Reality (AR) or "camera" view by default. This is a risky choice, especially coupled with the smaller icons, because a casual viewer might not notice the other icons or understand what they mean. As you can see the first casual AR experience is not particularly satisfying or useful (the labels block each other and are almost impossible to read), so the person might abandon the app after just one use and never discover the more useful List and Map views.

Also significant is that the active Camera tab is not shown to the customer, exacerbating the discovery issue. Not having the active tab shown makes it challenging to understand the relationship between the different views.

Interestingly, the older version of the Wikitude app used tabs that looked very different. The difference between this new version of Wikitude and the older version of the app shown in Figure 8.20 is germane to this discussion.

You can probably agree that the new visual design of the Wikitude app is an improvement over the previous version because it features slicker graphics. Tabs in the older version featured clunky icons and non-English text, and it took up a great deal of screen real estate. However, the older version was superior from the standpoint of learnability and usability. The default tab was List, followed by Map, and finally the Camera or AR tabs. This likely matches the typical use pattern of the tabs, which is an important plus. Another huge plus is that the active tab was visible and highlighted. This made understanding the app's IA a trivial task that could be taken for granted.

FIGURE 8.20: This is how the Tabs pattern was used in a previous version of the Wikitude app.

With plenty of space available for all three miniature icons, the designers could have easily incorporated the active tab and shown it next to two other tabs using a variation of the Android's "underline" treatment, as shown in the proposed wire-frame in Figure 8.21.

You'll probably agree that starting with the list view and using the blue under-line treatment to make the "you are here" tab obvious makes the app useful and usable.

FIGURE 8.21: This is a proposed wireframe of the Tabs treatment for the Wikitude app.

When and Where to Use It

Tabs are appropriate when the same data can be viewed in different ways, most commonly in a list or map, and the viewer needs to switch rapidly between the different views.

Tabs are also appropriate when the inventory can be segmented into various sort and filter buckets.

Why Use It

Tabs are easy to discover and are intuitive to use. Tabs borrow freely from the real world metaphor and represent an easy way to understand the contents of a collection or available view options.

Other Uses

Views do not necessarily refer to different representations of the data. Tabbed views can also be used to apply sophisticated combinations of filtering and sorting parameters to the same set of data and provide convenient entry points into vast, complex collections of items that cannot be searched easily. The classic example of this application of the Tabs pattern comes from the Google Play app (see Figure 8.22).

FIGURE 8.22: The Tabs pattern finds an effective application in the Google Play app.

Here the customer uses tabs to switch between Categories, Featured, Top Paid, Top Free, and so on. In the Play Store, tabs provide convenient self-explanatory entry points into a vast collection of apps by reducing the complexity into an interesting and quite manageable list of three visible categories of top results. These three lists are updated quite often and provide intrinsic entertainment value (think water cooler gossip such as "Guess what free app is most popular today?"), which make these tabs one of the most popular destinations on the Android platform. Google Play app offers an example of the scrolling tabs: allowing the customer to browse more than just the three tabs that can be seen on the screen at one time, by tapping on the partially hidden tab or via horizontal swiping. Although this method can show more than can be seen in the single screen view, this kind of display is also limited to a manageable size of 8 to 10 tabs; otherwise, swiping fatigue sets in and it becomes hard to discover all the various tab options.

Pet Shop Application

Tabs are versatile. In the Pet Shop app, you can use tabs to show different views of the search results—for example, List, Gallery, and Map views. (See Figure 8.23.)

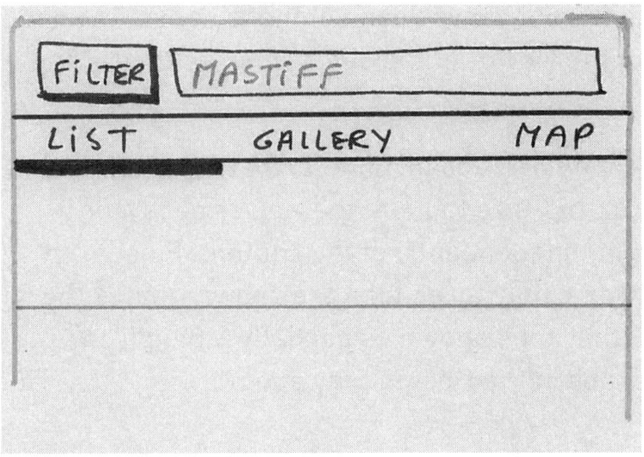

FIGURE 8.23: Tabs are used to show various views in the Pet Shop App.

But you can also use tabs to show two different ways to sort the results: Nearest, Newest, and Cheapest (see Figure 8.24).

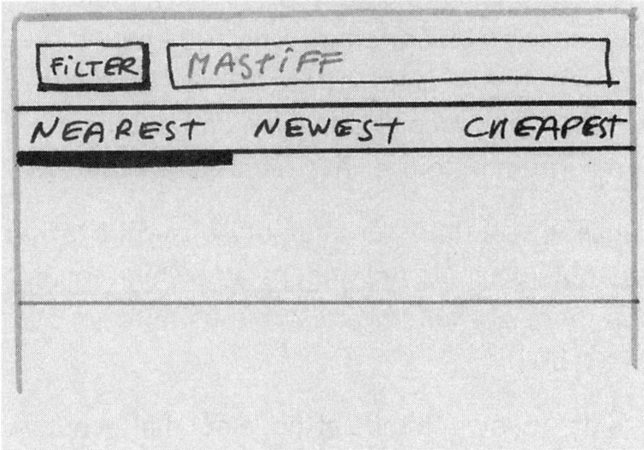

FIGURE 8.24: Here Tabs are used to show various sort orders in the Pet Shop App.

Which one should you use? As with any other design pattern, the answer is "It depends." That is, it depends on what fits best with what your customers are trying to do and what brings in the most business.

Tablet Apps

Tabs are fantastic for tablets and are vastly underused. Tablets have a lot more real estate, so the limit of only three tabs across the screen does not apply. Also, side-swiping the active tab to get to different views does not make sense. Instead, all 8 to 10 tabs can be freely shown to the viewer at the same time, as they are in the desktop web model.

You can even envision an effective Responsive Design of tabs: On larger tablets, the viewer sees all the 8 or 10 tabs. On mid-size tablets, the viewer sees 4 or 5 tabs, which can be scrolled left to right independently of the content. Finally, on mobile phones with the smallest screens, only three tabs are shown (that is the active tab is displayed fully with the other tabs shown as partially fading from view—refer to Figure 8.22 to see the tabs in the Google Play app).

! Caution

On small mobile devices, tabs take up precious screen real estate and therefore must be used with caution. Tabs don't need to have heavy visual elements, but reducing text into icons places the additional discovery burden on the customer. (Remember the Wikitude example from earlier in the chapter.) Typically, make tabs only as tall as needed to place the labeling text or small icon and enable accurate taps. However, no matter how small you make the tabs, they still need to be large enough to be tapped accurately. This means that even on the larger mobile phones, you can place only a maximum of four or five tabs across the row.

If you decide to use custom tabs, be sure to test them for usability with your target customers. The biggest issue is that people often do not understand which tab is "on." (This is a common problem when there are only two tabs but usually not a problem when there are three or more tabs.)

If there are only two tabs, or if viewers do not switch tabs often, a good alternative is a dedicated button to switch to a different view. A good example of this pattern is the Yelp app shown in Figure 8.25.

By providing a Map button instead of a tabbed view like Wikitude, Yelp removes the need to provide the tab bar and is free to devote full-screen real estate to the map view, which greatly improves the map view functionality and helps to create a more immersive experience (more on immersive UX in Chapter 13).

Does your customer actually need all those views on the same IA level? Maybe. Contrast Yelp with Wikitude: Although Yelp also has the AR function (called *Yelp Monocle*) and was the first app to introduce this feature, the Yelp Monocle AR view is deliberately separated from the List and Map views (see Figure 8.26).

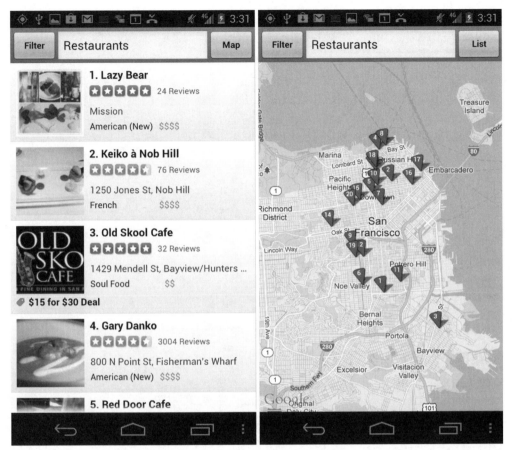

FIGURE 8.25: The Yelp app includes a dedicated Map/List button for switching views.

This is a clear signal that Yelp's designers do not place Monocle on the same IA level as the List and Map views, but instead invite the customer to engage with AR as an experiment or a diversion. In other words, AR is placed in a class by itself, away from the "working" List and Map views. Yelp offers a much more realistic and therefore more usable and less complicated implementation than the current Wikitude design that practically forces the viewer to engage with the AR as the first tab.

FIGURE 8.26: Yelp's AR view is separate from the List and Map views.

Related Patterns

8.2 Pattern: Refinement Page

CHAPTER 9

Avoiding Missing and Undesirable Results

When customers attempt to operate tiny mobile screens with a fat thumb, using only one hand, or while being jostled in the metro and eating a sandwich at the same time, mistakes are bound to happen. You must realize that those mistakes are not errors. They are a natural outcome of mobile computing, which takes place in a fast-paced, multitasking world. This makes avoiding missing and undesirable results a priority. The extent to which your app assists your customers in figuring out how to resolve the missing and undesirable results condition determines in large part their sense of satisfaction with your app, their brand loyalty, and whether they will recommend your app to their friends.

Essentially, recovery boils down to three essential elements:

- Telling the searcher that the system did not understand him

- Focusing on providing a way out

- Leveraging to the fullest extent the sensor and history information available in the mobile context of use

This seems like a straightforward strategy. Unfortunately, as you see in this chapter, most apps still struggle with relatively simple problems.

⊖ 9.1 Antipattern: Ignoring Visibility of System Status

On a small mobile device with its frequent user entry errors, it is tempting to bypass telling a customer about the problem and instead just take corrective action. This is generally better avoided by simply telling the customer "I did not understand," and thus it makes the first antipattern.

When and Where It Shows Up

This antipattern is fortunately not common. However, whenever the system takes some significant action on behalf of customers without telling them about it, this antipattern rears its ugly head.

Example

Here is an example from Yelp, an otherwise fine app, where the system takes unusual liberties in trying to guess what people are typing in. For example, in the image shown in Figure 9.1, the person is looking for sushi restaurants in Cupertino, a city in the heart of California's Silicon Valley. Unfortunately, the searcher mistypes the word *Cupertino* as *Coppertine*.

The results are sushi restaurants located in... *West Jordan*. Needless to say such results would be confusing to the extreme. The person may not even pay attention to the city marker, but instead attempt to call a restaurant to book a reservation or even try to navigate to it. Imagine the surprise! All this grief could be easily avoided if the system clearly stated that it did not understand the query rather than silently attempting to give its best guess instead.

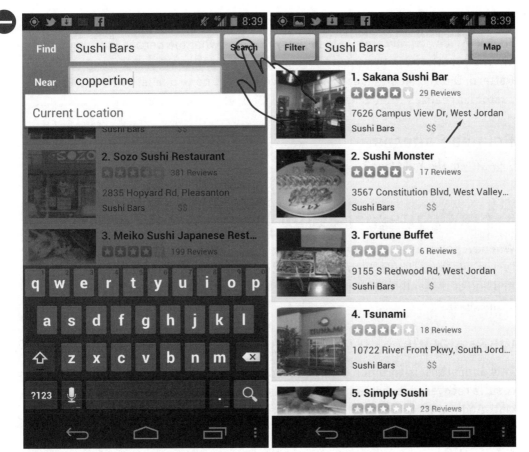

FIGURE 9.1: The Yelp app includes an example of the Ignoring Visibility of System Status antipattern.

Why Avoid It

The first strategy of the zero-results recovery pattern is to tell customers that the system did not understand them. Ignoring this fundamental strategy makes the entire system less trustworthy in the eyes of customers. If people are unaware that the app did not understand them, they might think that the app is malfunctioning or, worse still, violating the basic rules of logic and reason. Not stating that a misunderstanding occurred, violates the implicit agreement of the mental models people construct to operate complex machinery, so they get stuck. In the real world that means that customers feel frustrated but try again or—a worse scenario—move to a competitor's site or app and never come back.

Additional Considerations

To clarify, it's actually a great idea to try to guess what the person is trying to type in—many patterns in the chapter are devoted to doing exactly that (see the "9.4 Pattern: Did You Mean?" section). But if you *are* trying to guess, be sure that your customer realizes what is happening.

Related Patterns

9.4 Pattern: Did You Mean?

➖ 9.2 Antipattern: Lack of Interface Efficiency

Whenever a zero-results condition occurs, avoid the temptation to show an "error" state, particularly the one that requires an extra tap to get out of. Because missing or unwanted results are not an error, Lack of Interface Effieciency is an antipattern.

When and Where It Shows Up

This antipattern is common in apps that do not spend time designing missing results recovery mechanisms, but instead "blame the user" by showing the "official" Android error, and thus requiring an extra tap.

Example

Figure 9.2 shows an example from the Target app. When the keyword query is not recognized by the system, the app pops up an alert that says, Sorry, No Results Found Matching Your Request.

This alert inconveniences the customer in two important ways:

- It requires an extra tap to acknowledge the missing results condition.

- After the tap, the system navigates away from the Search Results tab and into the Shopping Basket tab.

This is both unexpected and inconvenient because the person looking for an item most likely mistyped it, so he needs to return to the Products tab again, which requires an unnecessary additional tap. Thus to recover from a common fat-finger mistype, the system requires two additional taps. This is an antipattern.

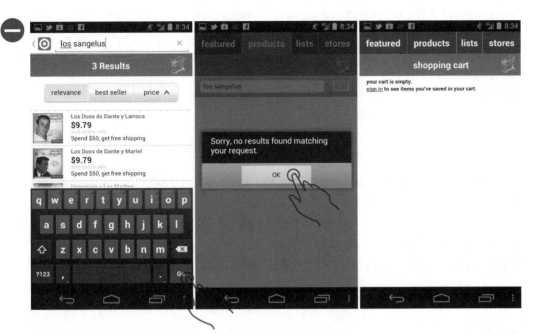

FIGURE 9.2: The Lack of Interface Efficiency antipattern is evident in the Target app.

The error dialog, plus additional tap required to get out of it, signal to the customer that he did something wrong and committed a sin…maybe even one of the unforgivable ones. If you recall that missing and unwanted results are both logical and natural outcomes of the mobile context of use on a tiny screen, it's easy to see that the extra tap acts as a sort of punishment added to an insult. You're hitting the customer's knuckles with a metaphorical ruler.

Additional Considerations

Recall that the second strategy in the missing or unwanted results recovery is focusing on providing a way out. The best way to do that is through an efficient, straightforward interface that does not treat any break down in communications with hostile dialogs. The dialogs and additional clicks they contain yank the person out of the state of flow, interrupting the search process abruptly.

Flow in search is important (read more on this later in the chapter), and it needs to be maintained at all costs. Dialogs that ask for additional taps and take the customer away from the task at hand (such as the Target app navigating to the Basket and away from the Products tab) break the flow. As Alan Cooper so eloquently said in his book *About Face* (2007, Wiley), dialog boxes "Stop the Proceedings with Idiocy." So although you must signal clearly to the customer that the misunderstanding has indeed occurred, you also must do it without breaking the sense of flow of the finding process. That's what keeps your customers going and keeps them engaged in the finding task.

Related Patterns

None

9.3 Antipattern: Useless Controls

All too frequently the controls provided by the interface on a zero-results screen are the same as those provided on a screen with actual results. This is an antipattern.

When and Where It Shows Up

The Useless Controls antipattern occurs on zero-results pages treated the same as those pages that have results.

Example

Unfortunately, this antipattern is quite common. Figure 9.3 is an example from TripAdvisor, where the customer accidentally fat-fingered the word "saucelito" (while trying to enter "sausalito"), and the system helpfully said, Sorry, but Nothing Matches Your Search. So far so good—the app avoids Antipattern 9.1 Ignoring Visibility of System Status. Unfortunately, the screen also displays the Filter Search Results link.

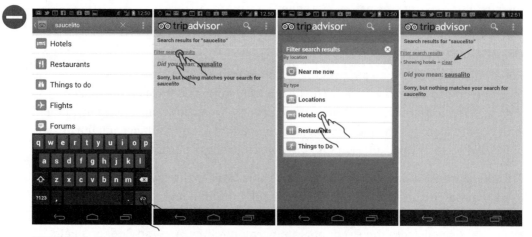

FIGURE 9.3: The TripAdvisor app demonstrates the Useless Controls antipattern.

Tapping the Filter Search Results link causes a filter dialog to display so that the customer can filter results by Destinations, Hotels, and so on. When the customer taps the Hotels button, he goes back to the same zero-results screen, but now the Hotels filter is applied. Of course, filtering zero results by hotels again produces zero results.

Why Avoid It

As you know from elementary algebra, zero divided by a number always yields zero. Zero results condition renders the filtering control useless. Actually, it's worse than useless because this filtering interaction drags the customer deeper into the quagmire of the zero-results search by having him go through the motions of actually filtering a nonexisting collection of zero items. Do not underestimate the customer confusion and damage to the experience that useless controls inflict. Useless controls are a clear antipattern. The zero-results situation requires that *all* the available controls focus on recovery and navigating somewhere useful.

Additional Considerations

None

Related Patterns

9.4 Pattern: Did You Mean?

9.4 Pattern: Did You Mean?

The simplest and arguably most powerful missing results recovery pattern is Did You Mean?—named after the Google search feature of the same name.

How It Works

When the customer types a keyword query that's not recognized by the system, the system offers a set of controlled vocabulary substitutions that are based on creative spelling of the original keyword query. The resulting list of keyword suggestions may or may not include the Did You Mean? title that Google uses, and it's not limited to a single suggestion. Given the number and creative quality of the misspellings possible on the mobile platform, the number of suggestions could also be correspondingly large.

Example

One creative implementation of the Did You Mean? pattern is done by the Booking.com app (see Figure 9.4).

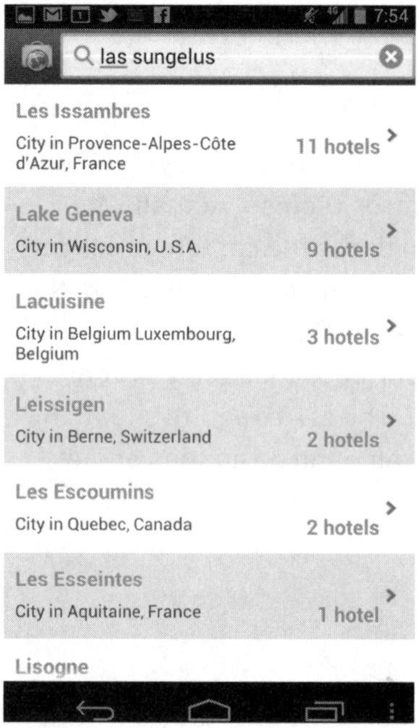

FIGURE 9.4: The Booking.com app includes a creative implementation of the Did You Mean? pattern.

Booking.com flips the entire search equation on its head by assuming from the beginning that you somehow mistyped the query. The result is a robust Did You Mean? interface that automatically suggests one or more controlled vocabulary substitutions for every keyword entry. Booking.com recognizes that the world is getting smaller, so you might just like to travel internationally, so it offers creative spelling suggestions from all over the world, as a kind of query prescreening. This is a nice example of the company branding expressed as a user interface feature.

One thing to note is that Booking.com adds the extra details (for example, Los Angeles, city in United States) which makes the extra click of the mandatory Did You Mean? query prescreening process worthwhile even for entries that are exactly correct in the initial spelling. After the city and country are positively identified, the added country meta data makes the final results for Hotels and Attractions reliable.

When and Where to Use It

At least one Did You Mean? suggestion is mandatory for any search suspected of keyword misspelling. Sometimes misspellings may be difficult to identify—a controlled vocabulary tuned to mobile-specific misspellings (such as one discussed in the "7.3 Pattern: Tap-Ahead" section) could be very helpful. Lastly, if you can't find any helpful Did You Mean? suggestions, you probably shouldn't suggest anything, just indicate zero results and use one of the other patterns in this chapter instead.

Why Use It

Misspellings caused by fat-fingering of the mobile device's tiny touch keyboard are common. Yet keyword queries are by far the most popular search methodology. This makes it imperative to implement the keyword query correction as a first line of defense against missing or unwanted results.

Other Uses

Did You Mean? is basically an Auto-Suggest pattern (refer to Chapter 7, "Search") that was implemented *after* the search has taken place. Given the tight bandwidth of most mobile devices, Did You Mean? is currently the more common pattern of the two. However, don't forget that almost anything that can be done after the search query is run can potentially also be implemented using Ajax on the front end *before* the query is run. Sometimes, Auto-Suggest is clearly superior to Did You Mean?, particularly in multi-field parametric queries, such as booking a flight. The name of the airport can be quite challenging to remember or guess, so a lookup with the Auto-Suggest based on the city name or airport code is essential. Ideally, at some point in the near future, as networks become faster, the entire process of two Auto-Suggest lookups (from and to airports) and the Did You Mean? feature in the results set (to suggest alternative departure times) becomes virtually seamless. In other words, the distinction we now draw between Auto-Suggest and Did You Mean? patterns will become even more blurred and eventually disappears entirely as you get instant feedback and suggestions for your query even as you view the results, much like Google Search now does on the web. Patterns such as Tap-Ahead (refer to Chapter 7) help show the way.

A related approach is to cache controlled vocabulary substitutions on the mobile device itself, much like the high-end paid language translation apps do currently. Mobile devices are actually full-featured computers. For specialized terms (like Pets or Airports) you can cache on the order of 100,000 or more Did You Mean? substitution terms locally and use the mobile device itself to run sophisticated regular-expression matching algorithms against this collection locally, instead of sending anything to the server. If you decide on using this approach, consider storing queries that cause the Did You Mean? function to activate in a log stored locally on the device, and periodically upload that log to the server, where it can be analyzed and used to improve suggestions in the future.

Pet Shop Application

Given the complex spelling of various pet breeds, it's easy to see how a Did You Mean? pattern can be useful in a Pet Shop app. However, it is tricky to see how to integrate Did You Mean? with various other refinement strategies. See "9.6 Pattern: Local Results" later in this chapter to see all the various missing and unwanted result strategies working together.

Tablet Apps

This pattern works the same way for tablets as it does for the smaller mobile phones.

! Caution

Did You Mean? is also sometimes called *controlled vocabulary substitution*. This means exactly what it says: The keywords suggested to the customer as part of the Did You Mean? recovery pattern come from some table in the database that contains the "good" or "allowed" keywords that form the controlled vocabulary. The quality of the substitution depends greatly on the quality of the controlled vocabulary, which, as discussed in Tap-Ahead (refer to Chapter 7) could be quite different on the mobile from its web counterpart. Not only can the queries be different, but the misspellings can also be different driven in part by the mobile keyboards. To make things even more interesting, there is anecdotal evidence that these misspellings depend in large part on ergonomics of the keyboard on a particular device. Therefore, iPhones, Android mobile phones with touch keyboards, Android devices with physical keyboards, Android tablets, and Blackberry phones may all give rise to different misspellings depending on the keyboard ergonomics of a particular device.

One idea might be to keep a specific database of the mobile-only misspellings and Did You Mean? suggestions. This enables you to better tailor the experience for mobile customers, without polluting the Did You Mean? database on the desktop web, which might include different misspellings. Whether you choose the approach with two split databases over the combined database depends wholly on your specific use case.

Related Patterns

7.2 Pattern: Auto-Complete and Auto-Suggest

7.3 Pattern: Tap-Ahead

9.5 Pattern: Partial Match

The second most important recovery strategy for missing results is Partial Match, which lets you recover from the missing results condition by omitting some search terms from the query.

How It Works

For queries with multiple keywords that return zero results, it is not always clear which keyword created the missing results condition. In this case, Partial Match Pattern can be used to recover. Partial Match reruns the query that consists of matching fewer keywords than were entered by the customer—that is, the system removes one or more of the keywords.

Example

One of the best examples in the industry comes from Amazon.com's mobile website (see Figure 9.5), which mimics the desktop in important respects.

FIGURE 9.5: Amazon.com's mobile website includes an excellent example of the Partial Match pattern.

The query Nike Ruskie Red produces zero results. The website clearly massages the condition. As part of the recovery strategy, Amazon.com helpfully does the search without the problem word "Ruskie" so that a results "appetizer" can be presented to the customer, which says in effect, "If you remove the problem keyword, you see more items like these." Strikeout coupled with the contrasting font color used for the problem keyword "Ruskie" creates a highly effective visual treatment that communicates how the Partial Match works.

Unfortunately the same cannot be said for the Amazon.com app. Although the mobile website works extremely well, the app does not provide any Partial Match recovery; it merely states that zero results have been found, as shown in Figure 9.6.

This forms an effective demonstration of the value of the Partial Match as a missing results recovery strategy.

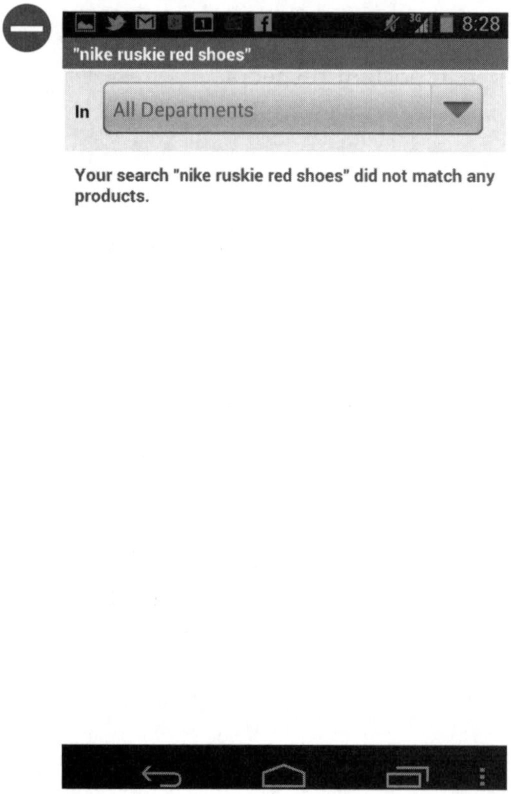

FIGURE 9.6: The Partial Match pattern is missing in the Amazon.com app.

When and Where to Use It

Partial Match is somewhat less important for mobile than it is for the desktop web. This is, of course, due to the difficulty of typing on the mobile, which results in slightly fewer average words per query. However, for some applications, e-commerce, and business apps, for example, the number of keywords entered on the mobile is quite close to that on the desktop. For those applications that employ long queries with multiple keywords, Partial Match forms the second most important recovery strategy after the Did You Mean? pattern discussed in the previous section.

Why Use It

Human beings are subject to a strong psychological effect called anchoring. *Anchoring* is a cognitive bias that refers to relying too heavily on one piece of information—in this instance, one query keyword. My book *Designing Search: UX Strategies for Ecommerce Success* (Wiley, 2011, ISBN: 978-0-470-94223-9) describes a situation in which the test participant was so convinced that the book she was looking for was titled *Harry Potter and The Sleepy Hollows* that she kept adding more and more information to the query (including misspelled author's name as "JK Rolins") over and over again, while the query kept returning zero results. The person in question anchored on the erroneous word "Sleepy" and no amount of failure would get her unstuck. She was so anchored on that term that at the end of the test, she erroneously concluded, "The store must not carry any Harry Potter books."

Her search behavior while on the system could be characterized as *churning*, what Peter Morville and Jeffrey Callender describe in their book *Search Patterns* (O'Reilly, 2011, ISBN: 978-0-596-80227-1) as running similar queries over and over while getting the same missing or unwanted results. The Partial Match pattern stops churning and anchoring behavioral antipatterns by showing clearly which keyword of the bunch caused the problem. The best implementations of Partial Match also return a sample "appetizer" (or the full search result) of the query without the extra keyword present to demonstrate the outcome of the query. Such implementations usually break the anchoring most effectively.

Other Uses

Sometimes it's hard to tell which keyword is an issue. In this case, Partial Match may provide multiple keyword combinations, with one or two keywords removed, each showing a small sample of items that would result from dropping the keyword(s). Usually these groups of results are sorted by the number of results retrieved from the query, highest first.

On a mobile device, Partial Match can also be effectively combined with the Local Results pattern, as shown in the "Pet Shop Application" section of "9.6 Pattern: Local Results."

Pet Shop Application

The complete implementation with the Local Results pattern is described in the "Pet Shop Application" section of "9.6 Pattern: Local Results."

Tablet Apps

Partial Match applies to tablets the same way as it does to mobile apps. Tablets have larger keyboards, thus the number of keywords per query is anecdotally similar to the number of keywords in desktop web applications. Therefore, the importance of Partial Match as a recovery mechanism is much higher. Simply showing a No Results Found message, as the Amazon.com app does (refer to Figure 9.6) is not resourceful. Besides, the No Results Found words would look so *lonely* on a giant tablet screen! Be sure to augment your tablet zero-results screen with one or more recovery strategies in this chapter.

! Caution

None

Related Patterns

9.6 Pattern: Local Results

7.3 Pattern: Tap-Ahead

9.6 Pattern: Local Results

When implementing missing results recovery on the mobile, take advantage of the local inventory of the results that can be retrieved using the on-board GPS.

How It Works

When a zero-results condition occurs, the system delivers some local results based on a partial match query or by providing a set of featured local results.

Example

The Target app store finder provides a simple example of the Local Results pattern. When the app first loads, it takes into account the customer's location via the on-board GPS. This location is used to display a map of nearby store locations even before the customer enters anything. This is a logical implementation because it is a safe assumption that 90 percent or more of use cases would find a local store. However, the app also provides a keyword search (presumably based on city name).

If the keyword entry cannot be successfully resolved into a city name, the app pops up an alert that requires tapping OK to dismiss it, which is an antipattern (refer to "9.2 Antipattern: Lack of Interface Efficiency"). However, after the dialog is dismissed, the apparent missing results "recovery" works well; the customer sees the local store results loaded in the previous step (see Figure 9.7).

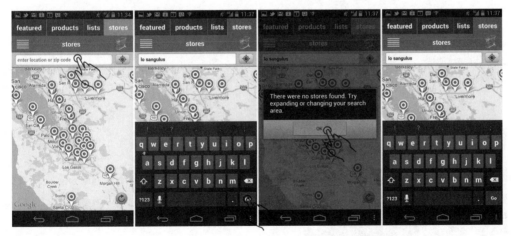

FIGURE 9.7: The Basic Local Results pattern assists with missing results recovery in the Target app.

When and Where to Use It

Mobile queries tend to be local. That's not the only use case, of course, but it's a common one. In the absence of clues from the customer as to the records the system needs to look for, or when executing a Partial Match pattern (refer to "9.5 Pattern: Partial Match") local results make an excellent bet.

Why Use It

When implementing zero-results screens, it is not enough to just avoid the anti-patterns. The best mobile experiences are "designed from zero"—that is, the prevention and recovery from zero results goes to the core of the way the search is implemented. Zero-results recovery based on the person's location determined via the built-in GPS is a true mobile feature not available on the desktop. Thus it constitutes a pure case of mobile-first advantage.

Other Uses

Local Results constitutes the "basic view" of the Parallel Architecture pattern (read more in Chapter 8, "Sorting and Filtering") and is the default search for most travel, reviews, and local shopping apps. Many of the excellent travel apps like Booking.com (see Figure 9.8) offer local results as a default browse tab—this one is labeled Around Me.

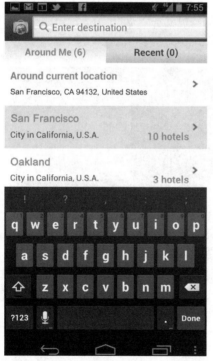

FIGURE 9.8: The Local Results pattern is used as the default Around Me tab in the Booking.com app.

Pet Shop Application

This is the moment you've been waiting for: Figure 9.9 shows all the missing and unwanted results recovery patterns (Did You Mean?, Partial Match, and Local Results) in one example running a query for "Russian Mastiff."

First on the page is the clear indicator that zero results have been found. Next are several Did You Mean? suggestions, derived from the popular keywords available on the site. Finally, because one of the keywords is correct and recognizable ("Mastiff"), the system performs a partial match, removing the problem keyword ("Russian") and redoing the search with the additional local component. This is a safe, useful, and resourceful assumption; most people like to buy their pets locally.

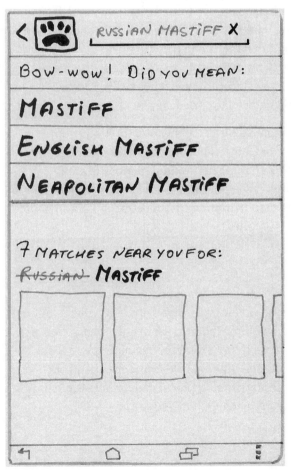

FIGURE 9.9: The Did You Mean?, Partial Match, and Local Results patterns work together in the Pet Shop app.

Sure enough, there are seven local "Mastiff" results to choose from, and a small sample displays on the screen so the customer has a good idea what these dogs actually look like. The customer can scan the results using the carousel or tap the 7 Matches Near You For link to redo the search with the keyword "Mastiff" and the location set to within a reasonable radius around the current GPS coordinates of the mobile device. With all three strategies working together, an excellent missing results recovery screen was created to rival that of the mighty Amazon.com mobile site! Recall how important this screen is to the success of people using your app, and spend a little extra time and care to make it work well for your customers.

Last but not least, I used a little Pet Shop humor ("Bow-wow!") to indicate zero results. This is highly recommended for most apps (other than Financials—refer to Chapter 12) as long as it is not the same funny line repeated over and over—that gets obnoxious in a hurry. (Recall the timeless scene from the first *Jurassic Park* in which the security system keeps saying "Uh uh uh! You didn't say the magic word!" over and over) Instead, have a few different well-chosen "wisecracks" picked at random to facilitate the missing results recovery.

Wisecracks are something at which Apple's Siri excels. Unfortunately, humor is where Android apps often come up short. Most Android apps still treat missing results as errors. Recall earlier in this chapter that mangled input is a natural and expected outcome of using a mobile device. Humor can be a wonderful tool to keep the search conversation going, despite momentary hiccups in human-mobile communication. And that's the important thing!

Tablet Apps

Although G3 and G4 tablets can both get a precise location, many tablets are purchased to run on Wi-Fi only and have no wireless connectivity. This makes obtaining a precise location of the device a problem. Tablets, especially large ones, are also less likely to be used in a local context (for example, navigating to a local Target store). Keep that in mind and rely on other recovery mechanisms such as Did You Mean? and Partial Match for tablets making local results available only if they can be obtained and make sense in the context of your app.

! Caution

See the preceding "Tablet Apps" section for this pattern.

Related Patterns

8.4 Pattern: Parallel Architecture

CHAPTER 10

Data Entry

Data entry on mobile devices is particularly tricky because of our fat fingers and the devices' smaller screens. There are literally hundreds of data entry patterns currently in use, and you could write an entire book specifically on this subject. Actually, my friend and mentor Luke Wroblewski did exactly that by writing an excellent book called *Web Form Design: Filling In the Blanks* (2008, Rosenfeld Media). In this chapter, rather than covering every conceivable input strategy currently in use, we concentrate on the trickiest aspects of Android forms, which people most often get wrong.

10.1 Pattern: Slider

Web page developers for years tried to popularize sliders and make them part of the standard HTML development toolkit. Many rejoiced that sliders came standard with the Android data entry widgets because, frankly, sliders are cool. Unfortunately, along with the sliders came a whole scope of issues that can sometimes be hard to pinpoint.

How It Works

Sliders come in two types: single and double. In addition to that, each type of slider can enable a continuous adjustment or have a set of predefined positions that customers can select from.

Example

To see how these two types of sliders work, it is instructive to compare two different apps: Trulia and Zillow. Trulia (see Figure 10.1) solves the price data entry problem using a dual slider.

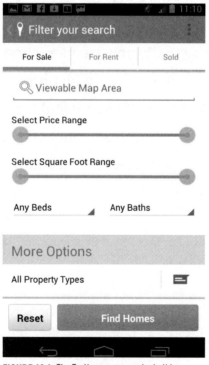

FIGURE 10.1: The Trulia app uses a dual slider.

On the other hand, Zillow offers the same price entry but has two single sliders (see Figure 10.2).

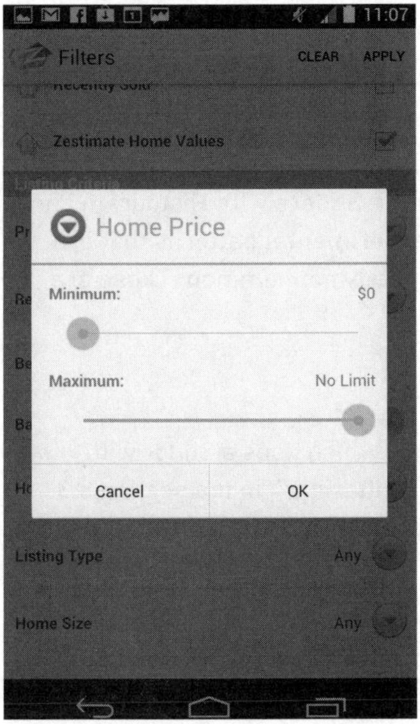

FIGURE 10.2: The Zillow app uses two single sliders.

The Trulia slider does not display the minimum and maximum range of values before the sliders are moved. Strictly speaking, neither does Zillow—instead it displays text $0 and No Maximum, which carries no information useful for the customer.

When and Where to Use It

Single sliders are intuitive for entering individual values. Dual sliders are great for ranges of values such as search filters and form values that include ranges.

Why Use It

Sliders are intuitive—they match what we call *affordance*—a quality of the control that makes it right for a particular task. In other words sliders just "feel right" for a particular task of dialing a value in a range. Sliders translate well from the physical world to the touch screen, where they look and feel great and are easy to manipulate without taking up a lot of space. Compare sliders to twist knobs, another physical control that can serve a similar function. In contrast to sliders, knobs do not translate as well to the touch screen, are hard to "turn," and usually take up more screen real estate than sliders. Dual sliders in particular are useful for bounding the interval within the specific range.

Other Uses

Single sliders with discrete values are interchangeable with Stepper controls, which are covered in the next section of this chapter. Sliders with Histogram and Slider based on Inventory Counts are two great experimental patterns that are variations of the standard slider that are unfortunately not common. These are described in the "Pet Shop Application" section.

Pet Shop Application

You can read about a Pet Shop dual slider example, which uses a Slider with Histogram design pattern, in Chapter 8, "Sorting and Filtering," in the section 8.3 "Pattern: Filter Strip." Sticking with the same example (that is, using sliders for entering price ranges on a search filter screen), can you use a dual slider to enter a price range without running into the issues listed in this pattern's "Caution" section? One way you can accomplish this is to use sliders with discrete values arranged according to inventory counts. Here's the explanation.

A typical price slider is arranged in a *linear pattern*, which means that an equal distance movement of the slider axis represents an equal *absolute change* in value. For example, in a five-position slider, the price can go from $0 to $100 in $20 increments, as shown in Figure 10.3 (numbers in gray are optional).

Select Price Range

FIGURE 10.3: Each mark on the axis represents an equal absolute change in value on a linear price Slider.

Although this is intuitive, this design makes it easy for your customers to shoot themselves in the foot, especially if the range is wide and the inventory is not equally distributed. For example, as explained in the "Caution" section, a customer might select a range of $40 to $60 in which the inventory is simply zero, not knowing that there is a whole wagonful of puppies in the $62–$65 range, literally—*For a Few Dollars More* (with apologies to Clint Eastwood and Western film lovers everywhere). This is where a Slider with Histogram (as shown in Figure 10.4) is helpful. The idea behind this experimental pattern is simple: The 50 to 100 pixels above the fixed position slider is the histogram representing the inventory in a particular section of the linear price range. Large numbers of dogs show a large bar, and smaller numbers show a proportionally smaller bar—that's it.

Select Price Range

FIGURE 10.4: A linear price Slider with Histogram pattern provides more information.

When using a Slider with Histogram pattern, you can still dial the part of the range with low inventory; it is just difficult to make that mistake accidentally because the inventory counts are clearly shown in the histogram. You can use the Slider with Histogram pattern where a standard discrete position Slider pattern would be used, taking up only a little more vertical space in exchange for a satisfying customer experience.

Another way to implement a Slider pattern without resorting to histograms, is to arrange the slider intervals based on the inventory counts. To do this, divide your entire inventory—say, 100 puppies—into five intervals, and you get 20 puppies per interval. Now scan the price range to figure out the price (rounded to the nearest dollar) that corresponds to the approximate inventory count of 20. Say the first 19 puppies cost between $0 and $60, (recall we assume no inventory in the $40 to $60 range). The second 21 puppies fall in the $61 to $65 range, and so on. Figure 10.5 is an example of what such a slider might look like (compare it with Figure 10.3).

Select Price Range

FIGURE 10.5: The alternative price slider is based on the inventory counts.

Which implementation should you choose? It depends on the task. Most people don't mind paying a few dollars outside the budget, but they absolutely *hate* getting zero results. If inventory is less than 20 items in a specific interval, that is not a satisfying result for most tasks, so you must use one of the other approaches to create a better experience. Both the Slider with Histogram and Slider Based on Inventory Counts patterns are far superior to the traditional Slider pattern. Breaking the interval by price is the flexible approach because it shows the distribution clearly, while never causing a zero-results condition. If the customer's price range is larger than that of a single 20-puppy interval, he can simply select a larger interval using the dual Slider. (Looking for the code for these experimental sliders? Visit the companion site for this book, http://androiddesignbook.com, for downloadable sample apps and source code.)

Tablet Apps

Sliders perform well in tablet apps. Make sure you heed the warnings in the "Caution" section; in particular, opt for the slider with discrete values instead of a continuous slider to ensure accuracy—it's harder to adjust a continuous slider accurately on a larger device. Take care of the device ergonomics, and avoid placing sliders in the middle of the screen. Instead, place the sliders near the top of the screen next to the right or left margin, optimized for one-handed operation with the thumb while the fingers hold on to the back of the tablet.

Depending on the design and purpose of your app, experiment by having two sets of sliders on the left and right sides of the screen to be adjusted by left and right hands, respectively. This would be especially interesting for apps such as music synthesizers. Finally, experiment with placing sliders vertically along the edge of the tablet (top to bottom) rather than horizontally from left to right, which is the easiest position to adjust precisely.

! Caution

Keep the following considerations in mind when using the Slider pattern:

- **Make sure reasonable values can be entered easily:** Kayak offers an example of a continuous dual slider for filtering hotel prices. To get a hotel in Los Angeles you can afford on a humble Mobile UX Design Consultant's salary, you must place the pegs right on top of one another (see Figure 10.6). This adjustment is anything *but* precise. For large ranges, consider using the version of the slider based on inventory counts, as described in the "Pet Shop Application" section.

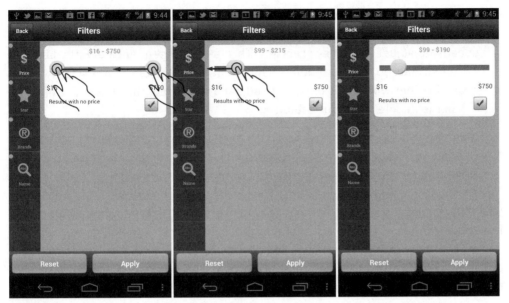

FIGURE 10.6: The continuous price slider fails to dial a reasonable hotel price in Los Angeles on the Kayak app.

- **Show the range:** Speaking of range, it's a great idea to show the actual range of prices available in the entire collection, as the Kayak app does in Figure 10.6 ($16 to $750) as opposed to using arbitrary numbers such as $0 and Max (refer to Figures 10.1 and 10.2). Neither Zillow nor Trulia show the true max and min associated with local home inventory. Imagine how useful these sliders would be if they actually said from the beginning that the range was from $476,000 to $3,234,700. Showing the range also helps avoid "dead zones," such as looking for a home in San Francisco priced less than $476,000, which yields zero results. Be aware of how filtering affects the inventory; it is best to rely on the range of the overall collection without the filters applied.

- **Don't cover the numbers:** While the customer adjusts the slider, the numbers should be shown above the pegs, where the fingers do not cover them. Placing the numbers below or to the side of the slider is not as ideal. The Kayak slider (refer to Figure 10.6) is good in this regard: The range is covered while the customer adjusts the slider, but the actual filter value is not, which is about the best you can do on a mobile device.

- **Opt for a slider with discrete positions:** Continuous sliders are sexy in principle, as is the idea of dialing the exact number and getting just the inventory you want. But the reality is that sliders are hard to adjust exactly—both in the physical world and on touch devices. That's why you almost never see a slider for volume adjustment on a stereo. Ironically, the larger the device, the harder it seems to adjust the slider precisely. This is the consequence of Fitts' Law: The time required for the action is dependent on the distance and size of the target. In other words, it's difficult to adjust a tiny peg of the slider 1/32 of an inch in the middle of a large tablet.

Regardless of the screen size, it is hard to adjust continuous sliders precisely while being bumped around in the metro or a taxi. (You have permission to refer to this thereafter as Nudelman's Law if you want.)

Continuous dual sliders also make it easy to over-constrain the range. To use the Pet Shop example, creating a continuous slider that enables the customer to dial a price of $45.50 to $46.10 yields a zero-results condition and does not serve the customer well. On the other hand, sliders with discrete positions (stops) are much easier to adjust. There is also less possibility of dialing a range that is too small.

Don't forget to consider the Slider with Histogram or Slider Based on Inventory Counts patterns. As described in the "Pet Shop Application" section, these experimental Slider pattern modifications offer a more *resourceful* user interface (the interface that helps customers act effectively and feel more capable) to avoid most of the pitfalls typically associated with Sliders.

Related Patterns

10.2 Pattern: Stepper

8.3 Pattern: Filter Strip

10.2 Pattern: Stepper

When the task calls for dialing a small whole number from 0 to 5, use a Stepper control.

How It Works

The Stepper is a native Android control that consists of the narrow text field flanked by minus and plus sign buttons.

Example

An excellent use of the Stepper is to dial the number of rooms and number of occupants on the search page from the Kayak app, as shown in Figure 10.7. Note the good use of reasonable defaults: Most people using the app are business travelers who travel alone and need only a single room so the Stepper controls are set to 1 in each case.

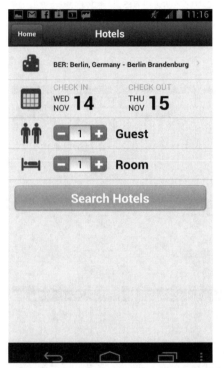

FIGURE 10.7: The Kayak app makes excellent use of the Stepper pattern.

When and Where to Use It

Whenever the customer needs to quickly enter a number between 0 and 5 where there is limited screen real estate, the Stepper pattern is the logical choice. A typical use is to adjust item counts during a purchase. Normally, the stepper value is submitted via a separate Submit button.

Why Use It

A stepper takes up little space and enables direct engagement. Like the Slider pattern discussed earlier in this chapter, the stepper can be manipulated directly on screen, without the need to pop up the keyboard or use additional layers, keyboards, or lightboxes. The Stepper pattern is self-explanatory and easy to use.

Other Uses

One interesting variation of the Stepper pattern is to use it as a row-level repeating UI control as a method to add items directly to the cart. This is how the Peapod app uses a customized Stepper control (see Figure 10.8). The first action is a simple Buy button. After the customer taps the Buy button, the item is added to the cart, and the button turns into a stepper.

FIGURE 10.8: The Peapod app uses Stepper as a row-level auto-submit control.

Whenever the customer taps a + or – button, the stepper control submits the new item count in the shopping basket to the server. This obviously slows things down a bit. However, most items come in only ones, twos, and threes, enabling this method to work well for shopping orders such as groceries, where total number of *products* purchased in a single shopping trip can be quite large, but individual *item* counts per product tend to be low.

One variation of the Stepper pattern as implemented by Peapod is the custom editable central text field. When the customer taps the central text field, the lightbox control launches so that the customer can edit the field directly (see Figure 10.9). So when the customer wants to buy those 99 Bottles of Beer for his Wall, he can do it easily by entering 99 using the lightbox keyboard.

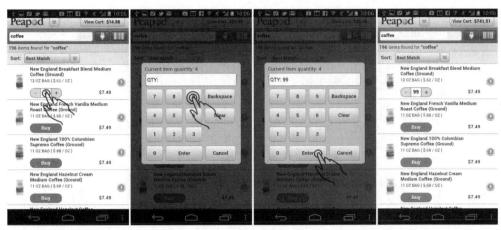

FIGURE 10.9: When the customer taps the center of the custom Stepper pattern implementation in the Peapod app, the numeric keyboard lightbox opens.

Without this customization (using the standard stepper control) the customer would have to tap the + button and submit the new count of beers to the server 98 times—as tedious as the famous song. (If you don't know the complete lyrics, check out http://99-bottles-of-beer.net/, which offers 1,500 different code variations in different programming languages with which to render the song.) Unfortunately, this unique customization is not discoverable—a better version of this is shown as part of our trusty Pet Shop app.

Another nice-to-have customization that allows steppers to enter larger numbers is press-and-hold: This action can be used to increment the count and accelerate the rate of increase with a longer button press. This is a very old pattern for devices using touch affordances. Of course this modification does not work for row-level Steppers that submit the value to the server with every tap.

Pet Shop Application

As mentioned, although Peapod's custom implementation of the Stepper pattern is highly effective, the discoverability of the editable text field in the center is quite low because the custom Peapod stepper looks similar to the standard Android stepper, which unfortunately does not have this direct data entry functionality.

Figure 10.10 is a hand-drawn wireframe of a custom version that takes care of this problem by moving the minus and plus buttons slightly further apart from the text field, making them round instead of square and enlarging the text field by a small amount. The resulting control looks customized but still carries all the correct stepper affordances; in other words, it's clear that the customer can tap – and + to increase and decrease the count in the middle.

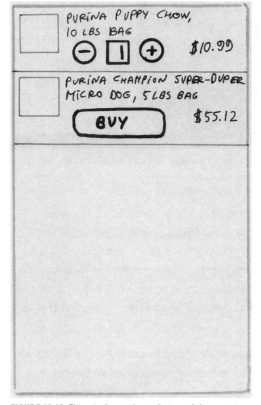

FIGURE 10.10: This wireframe shows the use of the custom Stepper pattern implementation as a row-level, auto-submit control in the Pet Shop app.

At the same time, this custom stepper also makes it more obvious that the customer can tap the middle text field directly to launch a "dial a number" lightbox. I also simplified the direct data entry lightbox from the one used by Peapod. Compare Figure 10.11 with Figure 10.9. Note the number in the textbox starts out highlighted for direct editing.

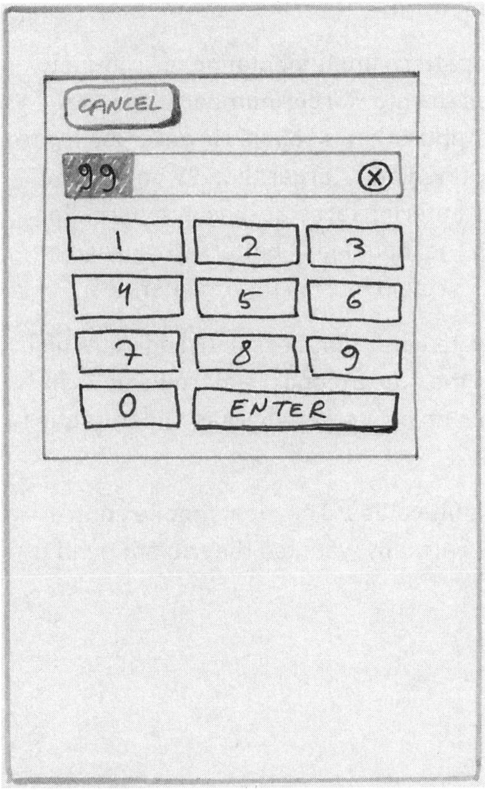

FIGURE 10.11: This is a simplified dial-a-number lightbox design.

The row-level arrow is omitted on the right in Figure 10.10, relying instead on people tapping the picture or item title to see the details. This is an example of the Tap Anywhere Android design principle discussed in Chapter 2, "What Makes Android Different."

Tablet Apps

Stepper works the same way for tablet implementations as it does for mobile phones. Remember that tablets are likely to display more rows than smaller mobile devices, so if you have a repeating in-row stepper control, you may want to make sure there is enough room to tap the –/+ buttons and also that the stepper control is positioned along the left or right margin of the device so that the customer can tap the control with a thumb while the rest of the fingers still hold on to the device.

! Caution

Do not attempt to use the standard Stepper pattern implementation for routinely entering numbers greater than 5. If you need to enter larger numbers, use the custom version described in the "Pet Shop Application" section. However, be aware that even the custom control does not accept numbers larger than 99 unless you make more modifications. Stepper is not for entering large ad-hoc numbers. For large numbers, instead use the textbox with the Input Mask pattern discussed in the "10.8 Pattern: Text Box with Input Mask" section later in this chapter.

You can use both a single slider and stepper to enter numbers 0 through 5. Which one you use depends on the exact nature of the app. Stepper is more flexible and can be used to enter numbers greater than 5 on occasion, whereas slider implies a bound (fixed) range of values.

Finally, ensure that steppers have good default values. The ideal mobile control is the one that has the value your customers already want, so they do not need to adjust it.

Related Patterns

10.1 Pattern: Slider

10.3 Pattern: Scrolling Calendar

When it's necessary to enter a date, most apps opt for a calendar that paginates by 1 or 2 months. This is traditional, but it's also tedious. A much slicker alternative is the Cadillac of Calendar Patterns: the Scrolling Calendar.

How It Works

The calendar control is the continuous scrolling ribbon of dates ordered in fixed columns by day of the week, with months separated by a thicker line.

Example

This pattern was designed by Alan Cooper and first described in his book *About Face: The Essentials of User Interface Design* (IDG Books, 1995—the updated third edition of this classic work on interaction design was published in 2007). However, it took more than 15 years for someone with the courage and vision to try it in a major application. That company was Kayak, which currently boasts one of the best implementations of this pattern available on the market. (See Figure 10.12.)

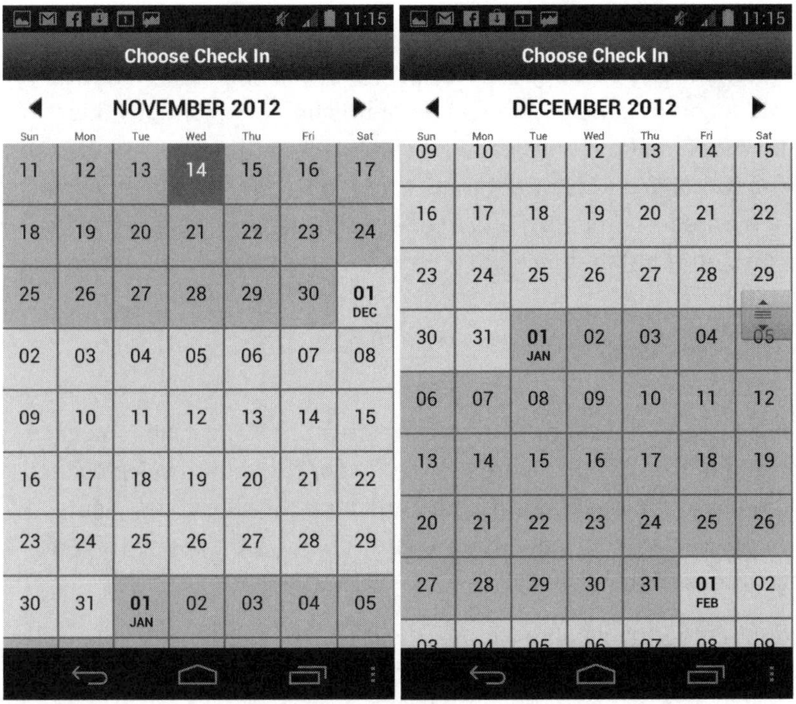

FIGURE 10.12: The Kayak app includes an excellent implementation of a scrolling calendar.

This calendar design takes advantage of the fact that although the number of days in a month changes from month to month (and sometimes year to year, as in February for leap years), the days of the week repeat continuously and unerringly—Monday through Sunday. Thus, the scrolling calendar consists of a ribbon that scrolls through seven columns representing repeating days of the week: Monday, Tuesday, Wednesday, and so on. Thus, even though the line separating the individual months is jagged, the dates flow through the days of the week in a continuous stream of numbers.

Note the additional "name of the month" feature is at the top of the page. As the month advances during scrolling, the name of the next month is partially displayed in the register. The customer can also paginate through the months in a more traditional manner by using the side arrows.

When and Where to Use It

Any time you need to enter a date that might be 1 or 2 weeks to several months in the future, the Scrolling Calendar is an excellent pattern to call upon.

Why Use It

Pagination of the calendar is a paradigm as old as the Gregorian calendar, which is *old*—430 years old as of the date of publication of this book. Essentially this includes the mental models of the physical wall calendar you flip pages on, which is not exactly the most modern implementation. As Alan Cooper eloquently points out in *About Face*, it's time for a modern and user-friendly design. I consider the Kayak mobile app implementation more usable even than its desktop website—an impressive accomplishment!

Other Uses

I used this Calendar pattern in a client app for a major U.S. retailer with one variation: The month on top of the scrolling calendar reflected the month *selected*, not the month displayed (as in the Kayak example). The default view is today's date. After the customer selects another date, the register displays the month of the new selection, regardless of the scroll state of the wheel.

To know which month currently displays in the scroll wheel, customers must rely on the three-letter abbreviation printed below the number 1 in the cell representing the first of the month. If you try this method, keep in mind that the range of available dates for that client goes to only approximately 2 months, which makes scrolling convenient, but also limited. If you have a longer time period to cover, you may want to stick with the Kayak implementation that dynamically shows the month displayed in the wheel as the calendar scrolls below.

Be aware that the Scrolling Calendar pattern is not just for dates; any continuous strip of timeline, including months, years, and most commonly hours of the day can be shown in a scrolling format that emphasizes the range between the two points.

Pet Shop Application

A wireframe for the Pet Shop app would not differ from the Kayak app, so it isn't reproduced here. However, feel free to indulge in a little practice by drawing this pattern yourself.

Tablet Apps

On a tablet device, a scrolling calendar would most likely be implemented as a popover. Remember to pop it up near the top half of the page and close to either the right or left side of the device for easy operation without taking your hands off the device. The top-right corner would likely be ideal for most people. Remember to plan for vertical and horizontal orientations of the tablet and accommodate both orientations in your popover placement and sizing.

Kayak does not currently do this well in the landscape (horizontal) orientation (see Figure 10.13). The lightbox that contains the calendar is located smack in the middle of the screen, making it necessary to let go of the edge of the tablet to scroll and select, even on the 7-inch device. This interaction is even more awkward for larger tablets.

FIGURE 10.13: The Kayak app places the scrolling calendar too far from the edge on tablets.

A much better approach would be to display the scrolling calendar to the right of the main form so that it's within easy reach of the thumb of the right hand.

! Caution

Remember to indicate the dates that are not available by graying them out. If it's important to your application, indicate today's date by putting the gray text "Today" in the cell under the number. Finally, don't forget that this pattern enables you to indicate gracefully a range of dates that spans a month or more, such as the length of stay, by highlighting the cells in the range for one-half a second just before the contents of the screen are sent to the server.

If your customer attempts to enter a date that is not allowed, make sure to message him before resetting the date to the first allowable date, *and keep the customer on the scrolling calendar screen*. Kayak currently does not do this, which makes the workflow confusing. Here is how the functionality currently works. Say the departure date is set to November 14 and the return date to November 18. The customer means to change the departure date to November 13, but mistakenly changes the return date instead. Obviously, this is Kayak, not Isaac Asimov's time-travel novel, so it's impossible to return before you arrive. However, instead of alerting the customer properly, Kayak *silently* resets the return date to be 1 day after the departure date—to November 15 in this case. (See Figure 10.14.)

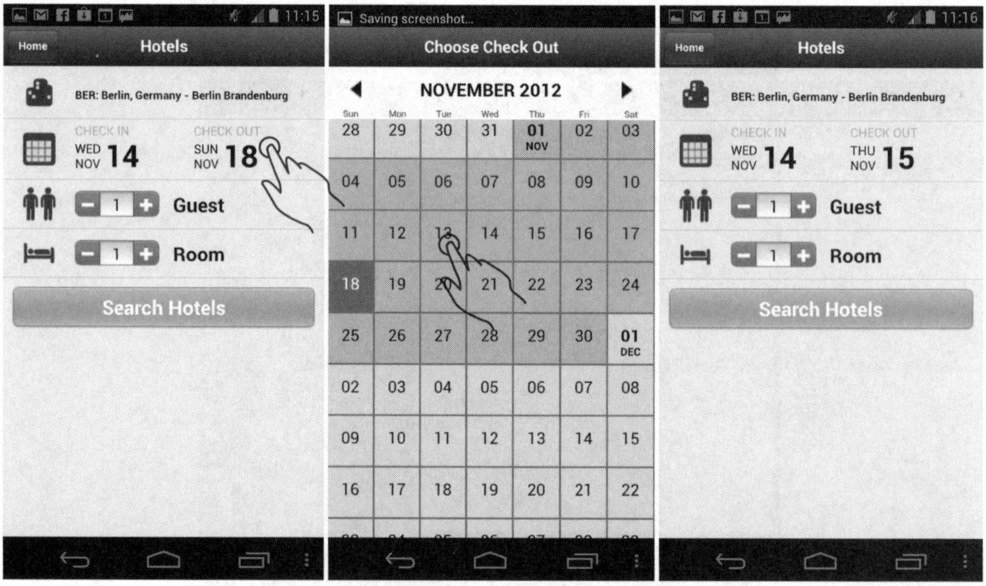

FIGURE 10.14: This is an antipattern. The Kayak app resets an incorrect end date silently, without any messaging.

Correct messaging around error conditions like this are the key to a successful, satisfying experience because date entry is unusually confusing to most people and causes multiple usability issues in user testing. Chapter 16, "Date Filters," in my *Designing Search* (2011, Wiley) book covers the details of data entry and display functionality.

10.4 Pattern: Date and Time Wheel

For complete flexibility in entering dates and times from an arbitrary length of time in the past or future, use the Date and Time Wheel pattern.

How It Works

When the customer needs to select a date, she taps the date or time field, which displays a lightbox of all the date components in a vertical picker format.

Example

Google Calendar provides a reference implementation example for this pattern (see Figure 10.15). Both date and time are shown on one line—a convenient and flexible, yet space-saving, arrangement. The presence of the Date and Time Wheel control is indicated with the Android's signature folded-corner triangle. Tapping the Date control brings up the Date Picker lightbox.

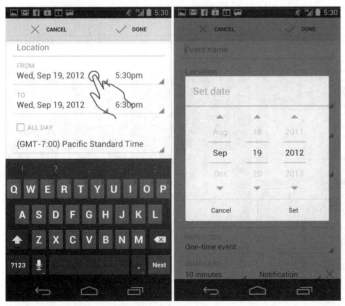

FIGURE 10.15: This is the Google Calendar app's reference implementation of the Date picker in the Date and Time Wheel pattern.

When the user taps the Time, the Time picker lightbox displays (see Figure 10.16).

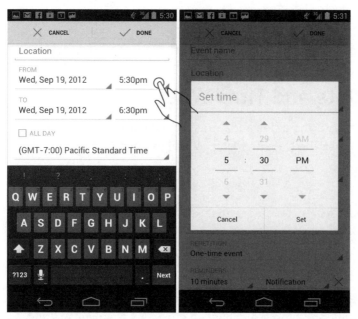

FIGURE 10.16: Here's the Time picker lightbox of the Date and Time Wheel in the Google Calendar app.

Android 2.3 and earlier wheel controls had to be adjusted by tapping the +/– buttons on the control or by tapping the textbox in the middle and editing the value directly via the keyboard (see Figure 10.17). There was no ability to use multitouch to "spin the wheel." If the person wanted to advance the values faster, she would hold down the appropriate + or – button.

FIGURE 10.17: This is the implementation of the Date and Time Wheel pattern in the Google Calendar app on Android 2.3.

Perhaps to use the back-door approach to get around the patent issues, or simply because the Apple's iOS Multitouch Picker control was cool, the Community Android developers created a similar after-market control called the Android Picker widget (http://code.google.com/p/android-wheel/), which is shown in Figure 10.18.

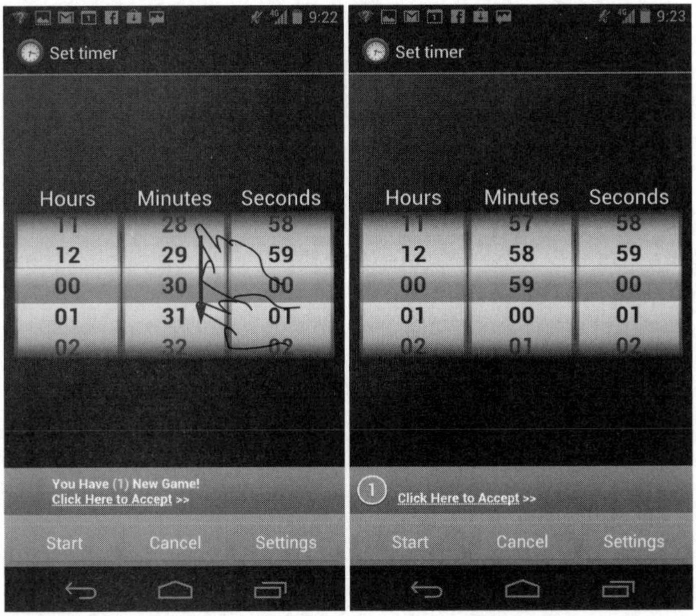

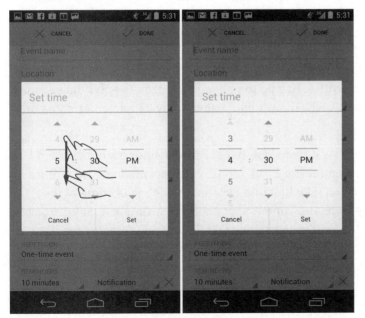

FIGURE 10.18: The Alarm Clock Xtreme app uses the Android Picker widget.

Starting with the Ice Cream Sandwich version (4.0) of the Android OS, the standard +/– wheel widget also includes the iOS Picker functionality for direct manipulation of the wheel's selection by swiping, as shown in Figure 10.19. For lack of a better name for this mode of adjustment, I call this swiping multitouch mode the Wheelie mode.

FIGURE 10.19: Android 4.0 includes the Wheelie mode in the Date and Time Picker lightbox, as shown in the Google Calendar app.

When and Where to Use It

Any time you need to enter an arbitrary date or time, this is the standard pattern to use. However, be aware that unlike the Scrolling Calendar pattern, this pattern does not show ranges. This pattern is also subpar for removing disallowed states (as described in this pattern's "Caution" section), so if the range of dates functionality is important to your app, you may want to look at the alternative patterns such as Scrolling Calendar or Dual Combo Wheels (see the "Pet Shop Application" section) for both the dates and time ranges entry.

Other uses of the wheel include any continuous ranges of strings 1 to approximately 20 characters. The only requirement to using the wheel control is that the values be obviously consecutive (numbers in order, seconds, months, and days of the week) and that the value can be displayed on a single line.

Why Use It

The Date and Time Wheel pattern is the standard pattern recommended by the Android OS, which enables complete flexibility.

Other Uses

Did you notice that the date and time entry required *separate sets* of multiple taps? If not, revisit Figures 10.15 and 10.16. With complete freedom comes complete responsibility. Entering dates and times like this is like playing Robinson Crusoe on a deserted island and having to make your shoes from scratch out of palm leaves and vegetable fibers. Yes, you don't have to go to the office every day where someone tells you what to do. But no, you can't get any Italian leather loafers in your size, and you have to chop all your own firewood. For most people, too much control is just as bad as too little. Having to use multiple taps to enter date ranges is *excessively* tedious, and it's why many people absolutely hate to enter dates on mobile devices.

So what's a designer to do? As it turns out, people rarely enter events years in advance. Most people do wait for the 2nd of January to start planning New Years' resolutions. (Shocking!) If you take that into account and let go of some of the flexibility of the separate Date and Time Wheel controls, a simple modification of the pattern takes care of the excessive taps issue. Figure 10.20 shows the iPhone screenshot of the date entry screen from the Pocket Informant app. The Combination Wheel control employed by the app consists of three parts: day of the week/date, hour, and minute.

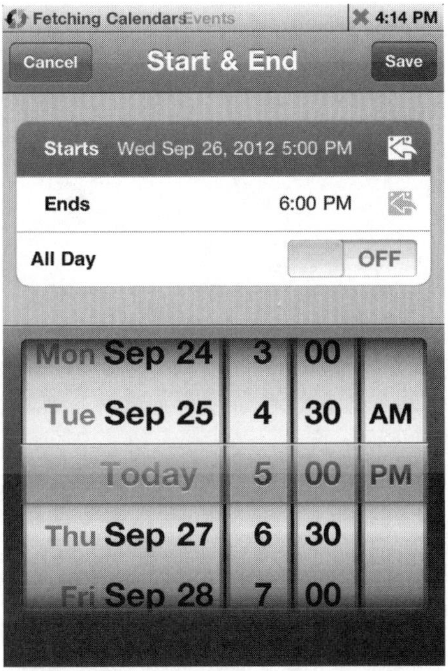

FIGURE 10.20: This Combination Wheel Date and Time Picker control is from the Pocket Informant iPhone app.

Having all the controls in the single wheel takes care of having multiple taps to open first the date wheel and then the time wheel, while still allowing plenty of flexibility. What you give up is the ability to easily pick anything, say, 10 years in advance. Most important, you can add a critical piece of data that was not available in the original design: *the day of the week*. Think of this Date and Time Wheel pattern implementation as a Robinson Crusoe who works remotely from his island on his MacBook Pro, ordering his Nike flip-flops from Zappos—all while sipping a frosty margarita.

Pet Shop Application

Imagine using a modified Date and Time Picker to book an appointment with a vet. Figure 10.21 shows a simple wireframe of what this implementation might look like on Android.

A combination control enables your customers to enter the date and time, and it shows the day of the week, which is important if you actually, you know, have stuff you need to do on certain days.

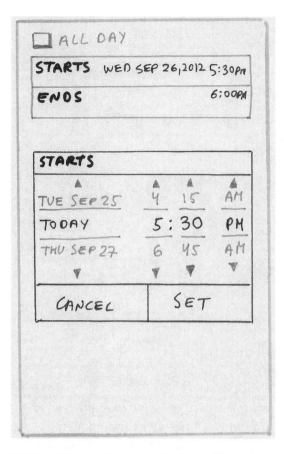

FIGURE 10.21: This wireframe shows a Combination Wheel Date and Time Picker for the Pet Shop app.

Another great solution would be an experimental pattern of dual composite sets of wheels for setting both from and to dates on the same screen, as shown in Figure 10.22.

This experimental pattern makes it much easier to enter dual dates on small tablets, allowing key information, such as the day of the week, to be clearly visible for both dates. This pattern would be particularly good for setting travel dates and meeting appointments on larger mobile devices and would cut down on taps while greatly improving clarity. You can even add the calculator of the event duration—handy indeed.

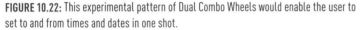

FIGURE 10.22: This experimental pattern of Dual Combo Wheels would enable the user to set to and from times and dates in one shot.

Tablet Apps

Although the Date and Time Wheel is a standard Android widget, it's actually hard to use on a tablet. Because the control is so tiny, it's difficult to swipe it using a normal-sized male thumb. This makes it hard to use the Wheelie mode of the Android control on a large device. (This is the same Fitts' Law at work on tablets once again as discussed previously in this chapter in the "10.1 Pattern: Slider" section.) On the other hand, the individual up/down arrows seem to work well on tablets, so the preferred implementation keeps the older Android 2.3 interaction mode, which disables the multitouch swipe control. To advance this wheel, the customer taps the +/− buttons (in this case the up/down arrows) and holds down the buttons to make the wheel advance faster (see Figure 10.23).

FIGURE 10.23: This is how the Calendar Android date wheel is implemented on a Galaxy Tab 2 7-inch tablet.

Note the scrolling calendar added to the right of the control. It's a nice touch; however, it does make it hard to adjust the control with the right hand, forcing left-handed use of the wheels or letting go of the device with the right hand. It might be better to swap the placement of the calendar and wheels to improve ergonomics.

Just as with the older Android 2.3 wheel implementations, you can adjust this control directly by tapping on the date and typing in the value using the keyboard (see Figure 10.24) .

FIGURE 10.24: You can type the date directly into the Calendar Android date wheel on a Galaxy Tab 2 7-inch tablet.

This interaction is currently reserved only for tablets and pre-4.0 Android devices. Directly typing the value into the wheel is not available on the Android 4.0 mobile phone implementations of this pattern.

! Caution

The general difficulty of swiping in the Wheelie mode was previously mentioned. This is inherent in the control, so there is little you can do about this, but you should generally be aware of the issue and look for the fix in later versions; increasing the height of the control, for example, would make it easier to use the swipe gesture.

On the other hand, there are some issues you do have control over. Watch out for the following:

- **Allow complete 360:** Be sure to allow each wheel to be turned completely around. For example, if the customer wants to select 40 minutes, she can start with 00 and go forward 01 to 40, or go reverse, 59 to 40. This helps avoid situations in which the larger number, such as 59, is selected and the customer must spin all the way back to 00 to reset the control. This also applies to on/ off switches such as AM/PM. The current implementation of the Calendar does not enable this (compare Figures 10.16 and 10.21).

- **Message entry errors and help recover:** If the business logic does not enable the customer to enter certain dates, the wheel control should message out the bad selection right on the date wheel lightbox itself without going to the main screen. Current behavior of the Google Calendar app in Android 4.0 is the anti-pattern; it accepts the bad *To* entry of September 19, which occurs before the *From* date of September 26, and then simply resets the *To* date back to the 26th without bothering to let the customer know that anything went wrong.

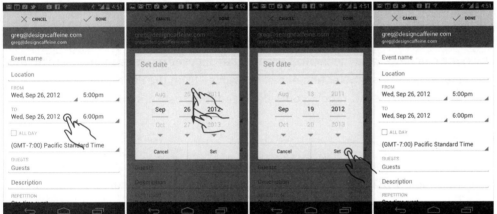

FIGURE 10.25: This is an antipattern; the Google Calendar app simply swallows date entry errors.

Remember: *Dates are tricky*. In the United States, many people make a mistake of entering the wrong AM/PM marker for appointments that span the noon hour, so they accidentally end up with meetings lasting more than 24 hours. If the person makes a mistake, the last thing you want to do is "swallow the error." As discussed in Chapter 9, "Avoiding Missing and Undesirable Results," you must let your customers know the correct system state and help them recover *as soon as possible*. That does not mean popping up errors all over the place. If you use a single composite wheel to enter both date and time (refer to Figures 10.20 and 10.21), you can *slowly* rotate the date dial to the next date using a smooth transition, while they are adjusting the time or AM/PM wheel. Most people notice the movement of the date wheel transition easily, and that is usually enough to cause them to self-correct. Another pattern that makes immediate error reporting and correction *in-situ* easy is the dual composite wheel experimental pattern shown in Figure 10.22, which would enable even more sophisticated business logic, such as excluding holidays or weekends, and, of course, not allowing the *To* date to be prior to the *From* date.

Related Patterns

10.2 Pattern: Stepper

10.5 Pattern: Drop Down

10.5 Pattern: Drop Down

Whenever you need an arbitrary value in the list of allowable values, Drop Down (also called Spinner or Select) is the natural pattern to call upon.

How It Works

When the customer taps the control's value, the popover box is launched with all the available Select values. Selecting one of the values from the list replaces the default value with the new value. Tapping outside the popover cancels the Select action.

Example

A typical example comes from the Trulia app (see Figure 10.26), where the customer uses the Drop Down to select a number of bedrooms and bathrooms in the search.

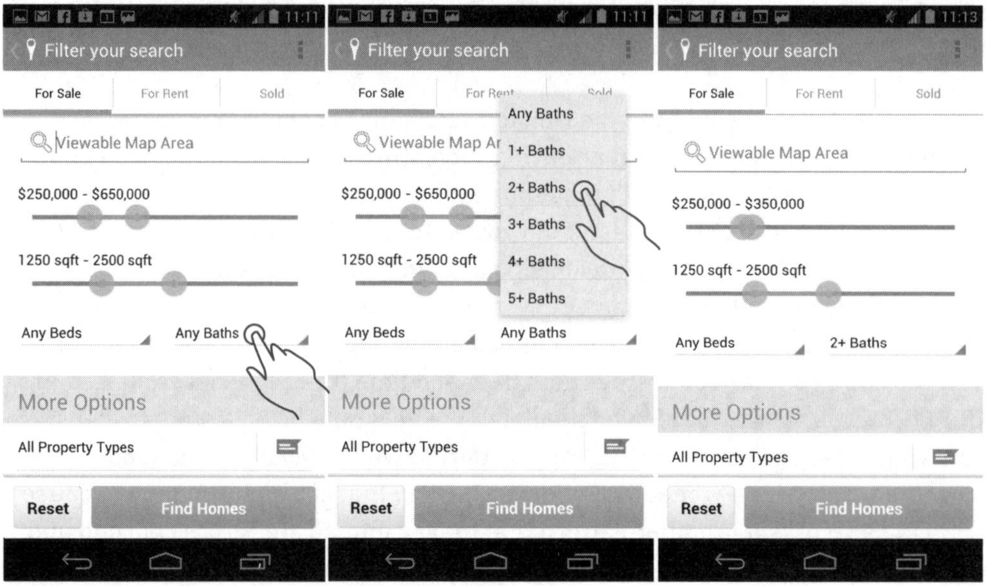

FIGURE 10.26: The Trulia app includes a typical Drop Down implementation.

Despite there being a reference Android 4.0 implementation, a different version of the control is offered by the competing app Zillow. Zillow's version of the Drop Down pattern (see Figure 10.27) looks a little different, but it performs the same function.

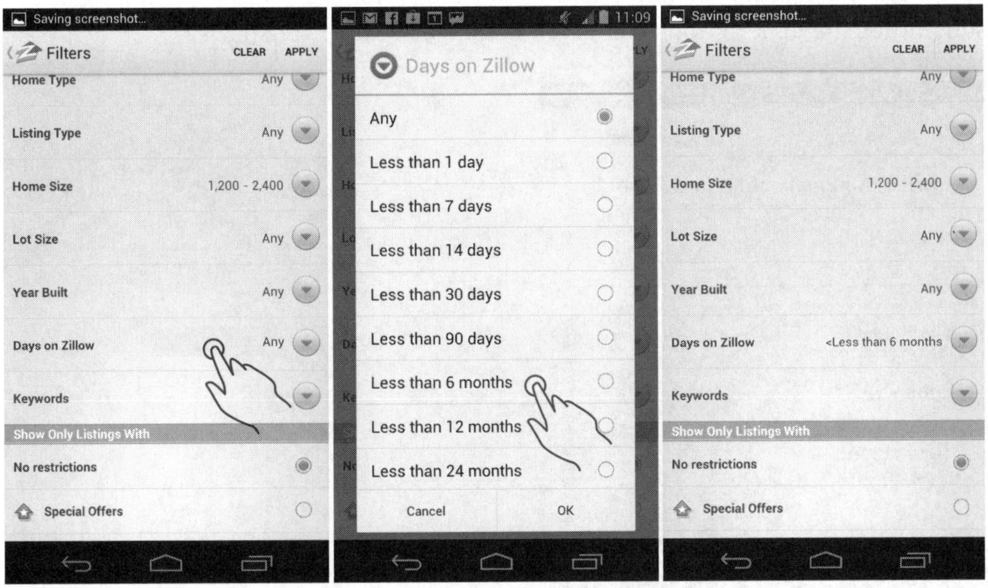

FIGURE 10.27: The Zillow app implements the Drop Down pattern a little differently.

When and Where to Use It

Use the Drop Down control any time you to need to allow the customer to pick from a list of 2 to 20 item sections that do not necessarily follow a specific sequence. The Drop Down control is also useful when the item text needs to wrap. You can use the Drop Down pattern in place of the radio buttons sequence.

Why Use It

The Drop Down pattern binds the value of a variable to a specific set of predefined values. It's intuitive to use and easy to operate.

Other Uses

One of the best applications of the Drop Down pattern is as an auto-suggest element. For example, in the Calendar app a modified Drop Down Combination control provides a space to type a few characters of the person's name and provides the auto-suggest list of contacts that can be selected to fill the Contacts field. This is an effective variation of the Drop Down pattern. Not only does the customer provide the criteria for selection (the initial few characters), but the Drop Down control is also augmented with a small thumbnail of the person's face to ensure slickness and positive identification. Where the picture is not available, the app substitutes the generic "portrait" thumbnail (see Figure 10.28).

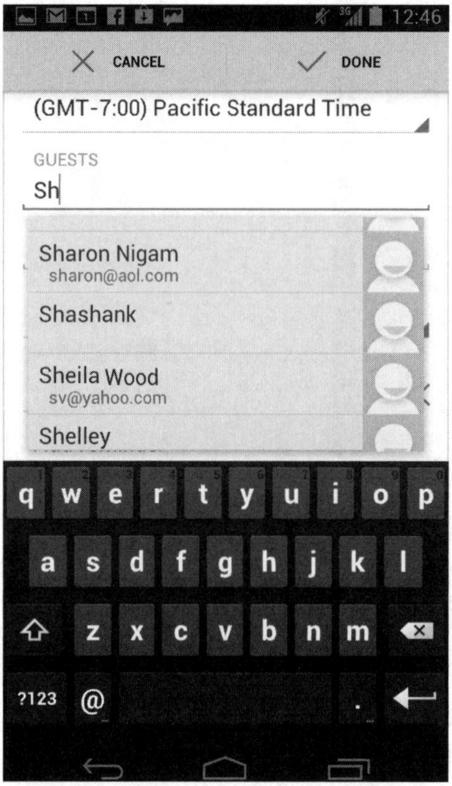

FIGURE 10.28: The Calendar app includes an implementation of the Drop Down Combination control with pictorial and text values.

See more about this pattern in the "10.9 Pattern: Textbox with Atomic Entities" section.

Pet Shop Application

For the Pet Shop app, I created a slick variation of the regular select with the pictorial content for selecting the type of pet the customer might want to search for (see Figure 10.29). For example, the customer can search for Cats, Dogs, Reptiles, and so on, and each type of pet is shown in the drop-down with a handy thumbnail. The thumbnails aren't drawn—this is just a wireframe. When you can practice drawing this pattern, and if you are so inclined, you can make your own pretty icons for each type of pet.

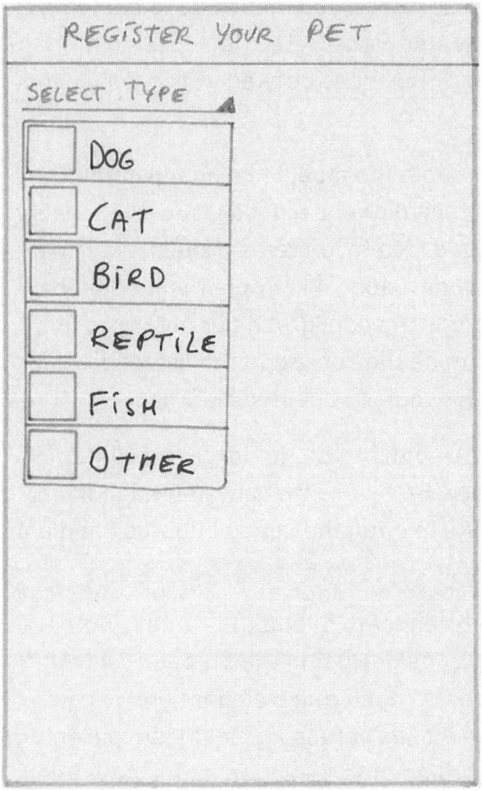

FIGURE 10.29: Check out the Pet Shop app implementation of the Select control with pictorial and text values.

Tablet Apps

On a tablet, this control works just the same way it does on the mobile device. As always, if you expect your customers to manipulate the control often, be sure to place the popover next to the right or left margin for easy reach with the thumb.

! Caution

Don't use this control for more than approximately 20 items. Scrolling through it might be tedious for your customers. A good alternative when you have more items might be the Dedicated Selection Page Pattern, or even a Combination Select control with a search function (refer to Figure 10.28).

Be aware that although the Drop Down pattern is quite versatile, it has limited styling options. For example, if you want to show a picture, text, and dollar value on a sky blue background, you must customize. When you do decide to customize, be aware of the competing pattern of a Dedicated Page Select (discussed in Chapter 12, "Mobile Banking") that discusses the differences between Drop Down and Dedicated Page patterns in more detail.

Some older or custom Drop Down controls enable the label to be on top of the control. Depending on the desired functionality, this might be a good idea—even though it is not strictly necessary. However, it is a good idea to implement the header that tells the customer what he is selecting in standard long-list implementations that are popped up in a lightbox window that darkens the background or obscures the page, to the point that makes it hard to determine the context of the selection. Use your own judgment here. Read more about this topic in Chapter 11, "Forms."

Be sure to create a good default value for the control—do not force the customer to make extra selections when not necessary. Be aware that any selection from the Drop Down involves at least two taps: one to open the control lightbox and one to select the item.

If the number of options is small (less than or equal to 5) and the value is numeric (such as number of passengers and rooms) consider using the Stepper control instead, as described in section 10.2. Stepper is much more elegant and is contextual in its selection. In other words, Stepper makes values inside it transparent to the customer and enables the customer to change the selection in the page itself, without popping a lightbox that can disorient him.

Related Patterns

12.2 Pattern: Dedicated Selection Page

10.4 Pattern: Date and Time Wheel

10.2 Pattern: Stepper

10.6 Pattern: Multiple Select

Sometimes, a single-select drop down does not quite measure up to a task that requires enabling the customer to select multiple values from a list. In those cases, Multiple Select is the logical answer.

How It Works

The drop-down version: When the customer taps the control, a popover box opens with a set of check boxes that enable the selection of multiple values from the list shown in the popover. After the selection is completed, the customer taps the Done button to apply the selections or the Cancel button to discard the changes and close the popover.

The gallery version: The customer can select one or more items from the gallery or list by tapping and holding down one of the items. This switches the gallery into a Multiple Select mode, which enables the customer to pick one or more items and perform actions on the selected items.

Example

One example of the Drop-Down Multiple Select control is to apply groups to a contact in the Contacts app (see Figure 10.30). The Multiple Select control enables the customer to pick one or more groups from the list and add the new contact to them.

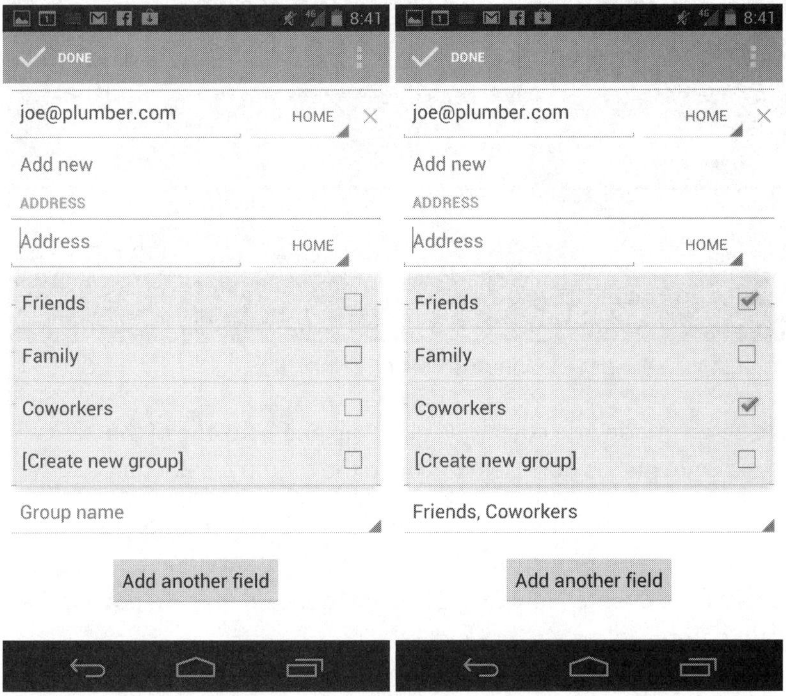

FIGURE 10.30: This is the Drop Down implementation of the Multiple Select pattern in the Contacts app.

When and Where to Use It

Any time your customers might want to select more than one item from a long list, the Multiple Select pattern is the way to go.

Why Use It

Sometimes, selecting one item at a time is simply too slow. The Multiple Select pattern provides flexibility. For bulk operations (such as selecting multiple spam messages in Gmail app) the Multiple Select pattern provides a graceful and usable alternative to performing the delete action one item at a time.

Other Uses

Another variation of this pattern is to select multiple items from the Gallery, such as the Android Photo Gallery app implementation. For the Gallery, multiple actions may include sharing, deleting, or even rotating the selected photos (as shown in Figure 10.31).

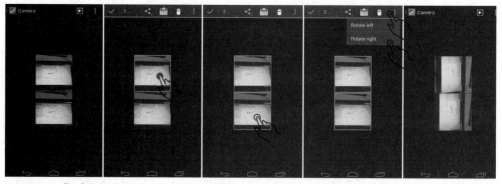

FIGURE 10.31: The Gallery variation of the Multiple Select pattern is in use in the Photo Gallery app.

To enter the multiple select mode, the customer must tap and hold one of the items in the gallery. After that a single tap is sufficient to add to the multiple selection.

For the Drop Down version of a Multiple Select control, one interesting feature is the ability to add a new element. Figure 10.32 shows the Contacts app again, but this time the customer taps Create New Group, so she gets a separate pop-up for entering a new group name.

Pet Shop Application

One idea of demonstrating the Multiple Select pattern in the Pet Shop app is to use it to pick pets you are interested in to add them to a wishlist for possible future contact with the seller. Actually, this is straightforward to wireframe, as shown in Figure 10.33.

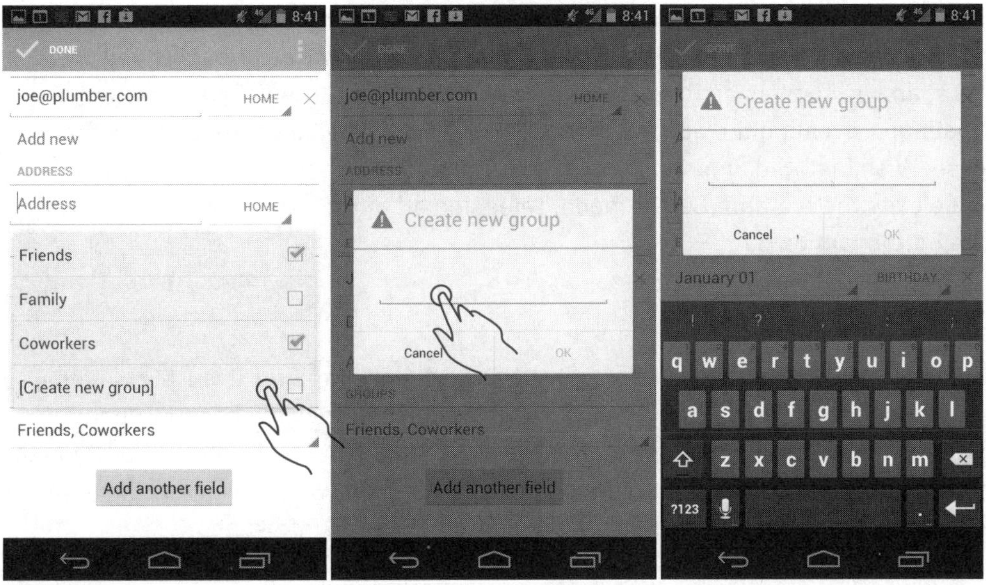

FIGURE 10.32: Another variation of the Drop Down Multiple Select pattern is being able to add a new value in the Contact app.

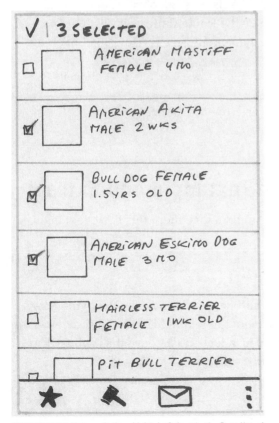

FIGURE 10.33: Using a Gallery Multiple Select in the Pets list of the Pet Shop app might look like this.

This Pet Shop functionality works similar to the way Gmail app Multiple Select works. In contrast to the Photo Gallery app, the check boxes are shown all the time, and as soon as an item is selected, the contextual menu on the bottom floats up, revealing the actions that can be taken on the selected items (in this case, Wish List, Bid, and Contact Seller). Tapping the check box on the top of the page exits the Multiple Select mode, releasing all of the selections and removing the bottom action bar.

Tablet Apps

Customers using tablets presumably have a bit more time for leisurely contemplation and to perform maintenance tasks; therefore, they might draw upon a multiple select even sooner than their mobile counterparts. It's not uncommon to provide Multiple Select functionality for tablets, and it works great to accommodate those urges to tidy up a gallery or add some group information.

! Caution

This option tends to be less discoverable—that is, it is not obvious how to enter a Multiple Select mode in some apps, such as the Photo Gallery. If this function is essential for your app, be sure to provide a quick tutorial overlay. See Chapter 5, "Welcome Experience," for ideas.

Related Patterns

10.5 Pattern: Drop Down

10.7 Pattern: Free-Form Text Input and Extract

Whenever customers need to enter random text, give them this basic text input pattern.

How It Works

The customer taps the text field, and the soft keyboard comes up from the bottom of the page, while the focus is placed into the text field. If the field is below the fold, the form scrolls so that the text field is in focus just above the keyboard. The customer enters a desired value for the text field via the keyboard.

Example

One example of this pattern is the Contacts app (see Figure 10.34) . The Contacts app appears with the soft keyboard already in place, ready to enter values.

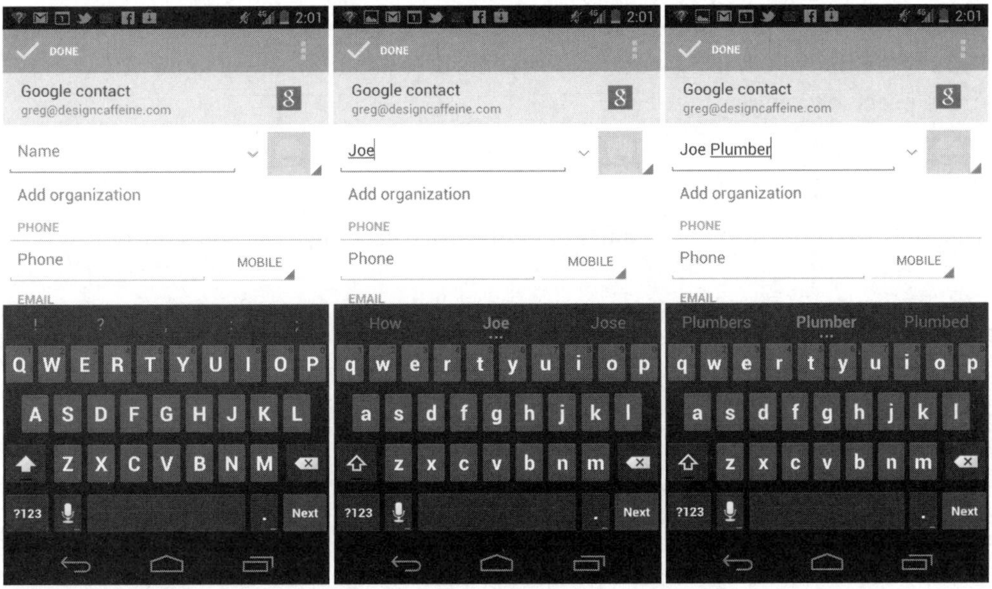

FIGURE 10.34: This typical textbox implementation is in the Contacts app.

Turning the mobile device horizontally while maintaining focus on the textbox launches an important variation of this pattern: the Extract Text Input (see Figure 10.35).

FIGURE 10.35: The Extract Text Input variation is displayed when a device is in the horizontal orientation.

When and Where to Use It

Use this pattern any time you need a free-form text input, such as the name, description, status update, and the like. The Extract version of the pattern is particularly useful for longer messages, such as an e-mail message body, because it make the input field larger; so the customer can see two to three lines of text at the same time, and has a larger soft keyboard, which for many people allows for more precise typing.

Why Use It

This pattern is the standard control to enter text via a keyboard on Android.

Other Uses

Starting with the Nexus phone in the Android 4.0 OS, any form field that accepts text also accepts voice input. Tapping the microphone button on the keyboard (or on some models, swiping across the keyboard horizontally) brings up the voice input mode. The keyboard swiping method is less discoverable, so fewer people know about it.

This voice input function is handy, especially for status updates or longer fields like an e-mail body. Unfortunately, it's still somewhat incomplete because the customer needs to tap the Next button to send the update or move onto the next field. (See Chapter 7 for discussion of the differences between standard voice input and the true convenience of digital assistants such as Siri.)

Pet Shop Application

This implementation is straightforward, so I did not create a wireframe for it; however, you can practice on your own.

This is a good place to mention the one recommended departure from the sticky-note wireframe style of the previous chapters: not to draw various keyboards in the sticky-note Pet Shop app wireframes. Drawing keyboards is tedious and time consuming, and it yields no tangible benefit in user testing. Instead, do what works: Simply print and cut out various keyboards and glue them onto smaller sticky notes using a glue stick or tape. Once created, you can place these sticky-note keyboards on top of the wireframed screen to simulate the keyboard coming up from the bottom. Then these paper keyboards can be moved down as needed to give the participants the feel of a complete sequence of filling out fields on the page. The sticky-note keyboards can be reused on other screens and projects, as needed.

Tablet Apps

On tablet devices, there is far more real estate for the form, so the form never enters the "extract" mode. Typically, the form just scrolls up to where the target field comes up just above the keyboard, regardless of the orientation (see Figure 10.36).

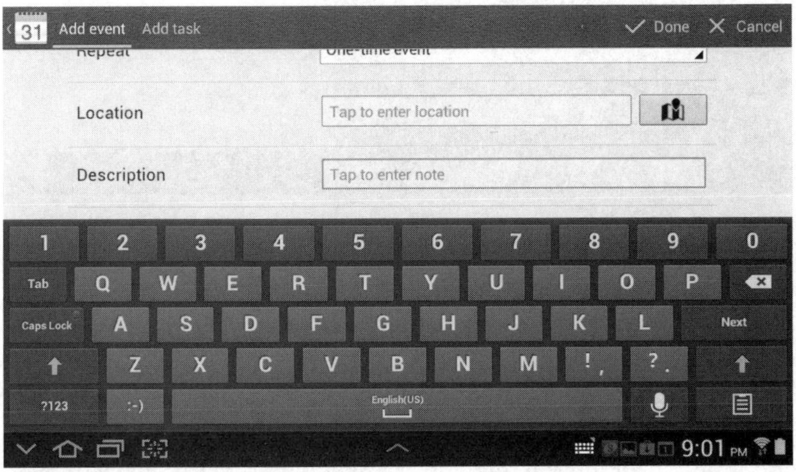

FIGURE 10.36: This is how a textbox implementation looks in the Calendar app on the 7-inch Galaxy Tab 2 running Android 4.0.

Unfortunately, flipping the tablet from vertical to horizontal orientation and back again sometimes closes the keyboard. This is definitely a bug; the keyboard should stay open until dismissed or until the customer taps Done or goes to the next field. See more about keyboards in the following "Caution" section.

❗ Caution

The only button that the Android toolkit provides for navigating between the text fields of the form is the Next button, which changes to Done when the last *consecutive* text field is traversed (see Figure 10.37). This works okay *as long as your text form fields are a single line.*

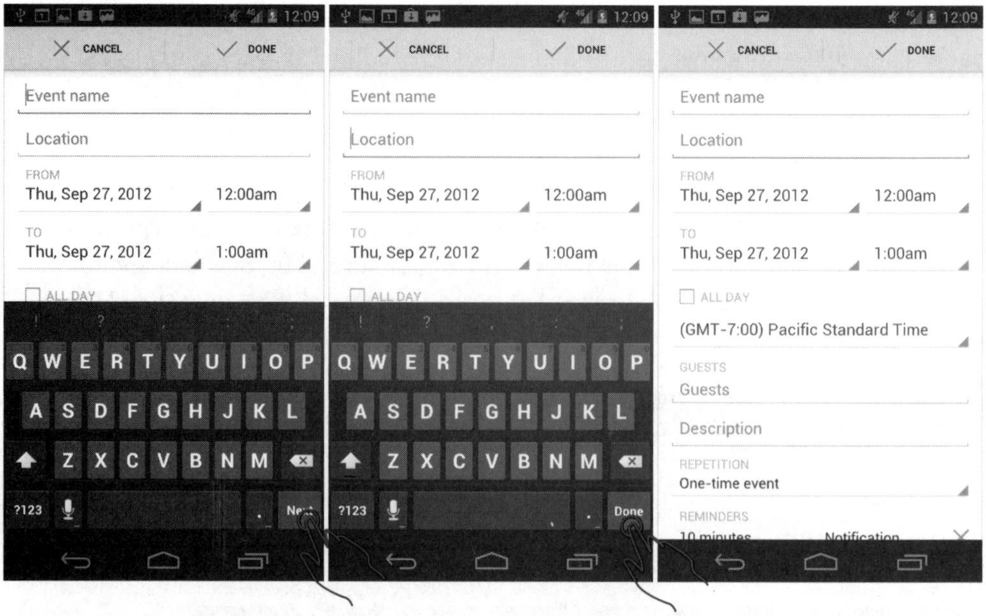

FIGURE 10.37: The Next and Done buttons appear on textboxes in the Calendar app.

Unfortunately, things break down when your form includes a multiline text field, such as a long description, which omits the Next or Done button in favor of the Enter key. If you happen to be in the multiline field, the keyboard remains on the screen until the form is submitted. So in effect, for the entire remainder of the form, the customer is left scrolling the form up and down with her finger to move into the next field in a small fraction of what remains of the screen real estate above the keyboard, while the system focus remains on the multiline description field. This is an antipattern, as shown in Figure 10.38.

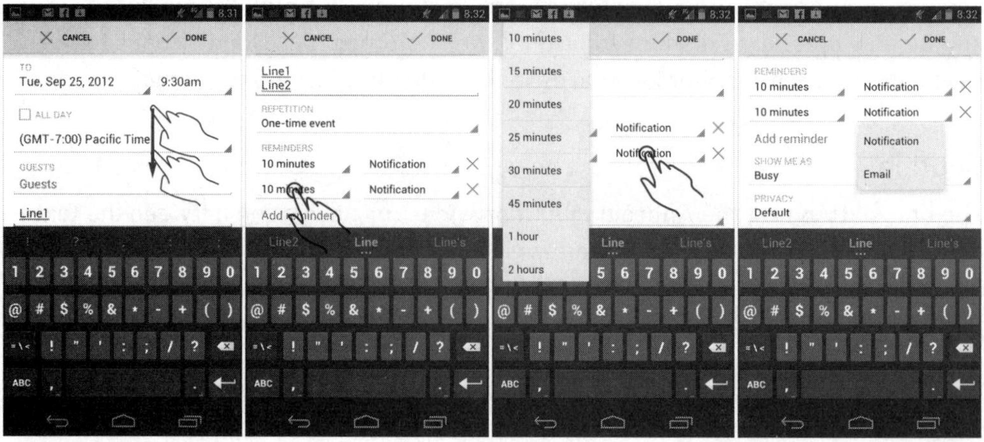

FIGURE 10.38: In Google Calendar, the keyboard for a long description field remains on the screen even while the focus shifts to other controls.

You might say that the standard behavior is to use the hardware Back button to exit the keyboard while in the multiline description field. That is true; this function is available. However, the discoverability to use the hardware Back button to get rid of the keyboard is low. This function is counterintuitive because the multiline entry field looks exactly the same as the single-line field, minus that critical Next/Done button. To add to the confusion, this same multiline text field has an explicit Done button when viewed in a horizontal format (see Figure 10.39). The Landscape Done button moves to the next field, whereas the Portrait Done button actually submits the form!

FIGURE 10.39: In Google Calendar, Done buttons in landscape and portrait perform diametrically opposite functions.

The absence of the way to exit the long field explicitly (that is, the Next/Done button) means that you must be careful with the keyboard behavior, especially in portrait mode on smaller mobile devices where the keyboard obscures most of the rest of the screen. For this reason, *the keyboard should collapse as soon as the customer taps any other form field, regardless of type.*

You should also avoid using the keyboard Done button as a "form submit" that sends the form to the server. At present, the prevalent paradigm is for the Done button in the keyboard space to dismiss the keyboard. If you do use the keyboard button to submit the form, at the least, label the button with a distinct label, such as Go, and make sure the form submits only from the last field in the form. If space permits, it is best to provide a separate dedicated button for the Submit function.

One last note of caution: When your form or lightbox contains a single text field, strongly consider launching the form with the keyboard already in place to save your customers an extra tap. For example, in the Contacts app, when creating the new group, the lightbox with the single entry field is shown without the keyboard. The customer must tap the field to launch the keyboard. Another weird thing happens after the customer taps the OK button on the lightbox: The focus shifts upward to the Address field (see Figure 10.40).

FIGURE 10.40: In the Contacts app, the single text entry lightbox is launched without the keyboard and focus shifts randomly.

A better interaction would be to launch the lightbox with the keyboard already in place (refer to the second screen from the left in Figure 10.40) and shift the focus to the groups Multiple Select field after the new group name has been selected.

Related Patterns

10.8 Pattern: Textbox with Input Mask

10.9 Pattern: Textbox with Atomic Entities

12.5 Pattern: Wizard Flow with Form

10.8 Pattern: Textbox with Input Mask

Sometimes, we know more about the specific content of the text field, for example, that it contains an e-mail. In those cases, use the textbox with the Input Mask pattern.

How It Works

When the field accepts only specific data, such as e-mail, phone number, Social Security number, or ZIP code, the system can provide the right kind of keyboard to facilitate data entry. Also, depending on the location of the field label, it can optionally show the input mask inside the field.

Example

When adding a new contact, the Name field provides a "Name" in-field label shown in a lighter gray font inside the entry field (see Figure 10.41). The in-field label remains in place until the customer starts typing.

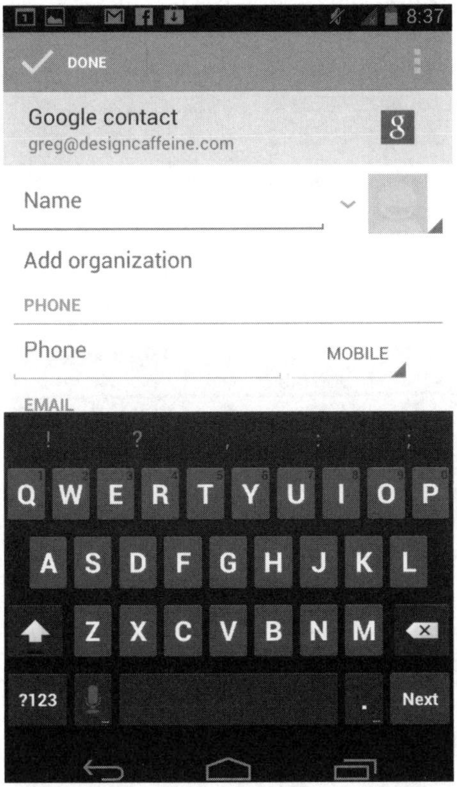

FIGURE 10.41: The Name field label acts as a kind of input mask.

You can think of the Name label as a kind of input mask. Some people might disagree, saying don't mix apples and oranges; they would argue that, strictly speaking, Name is not an input mask, but a field label—and that's okay. On a mobile device, with the tight screen space available, the differences between the functions of input masks and labels are, at best, a bit blurred, if not entirely muddled.

Despite being open to all sorts of characters, the Name field has some restrictions, such as field length, which is always limited to some practical number (for example, 100 or 255 characters) and disallows the "enter" character (also called "new line" or "carriage return"). This makes the Name field different, from, say the Notes field. The Notes field enables up to 1 Mb of text data and accepts a press of the Enter key to add a new line to the text. In this way, it is useful to think of anything written inside a field that also happens to shape the customer's expectations of the format and allowable values as a kind of input mask.

So what is the Input Mask pattern good for? One useful feature is to facilitate the correct data entry by launching the right kind of keyboard. For example, notice in Figure 10.42 how the soft keyboard changes accordingly when the Name, Address, Phone Number, and E-mail fields are entered; that is the implicit input mask at work.

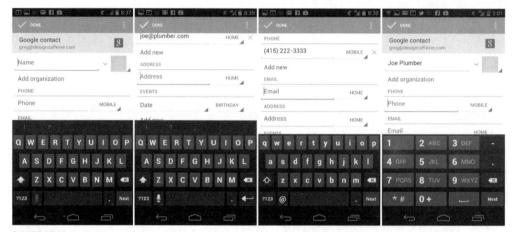

FIGURE 10.42: Implicit input masks launch various types of mobile soft keyboards.

The Name and Address both provide voice input capabilities, whereas Email provides a handy @ sign (which can be hard to find on the standard keyboard). Most of the new Android 4.0-style minimalist native form fields are going to be limited to this type of implicit input mask—that is, an input mask that is logically implied by the label.

For fields that have explicit persistent labels, a better customer experience is possible—one that provides a label and a separate input mask. For example, in the registration form for Kayak, as shown in Figure 10.43, the Email label to the left of the field also enables the input mask you@example.com that explicitly shows the expected field value format and the "required" input mask for the Password field.

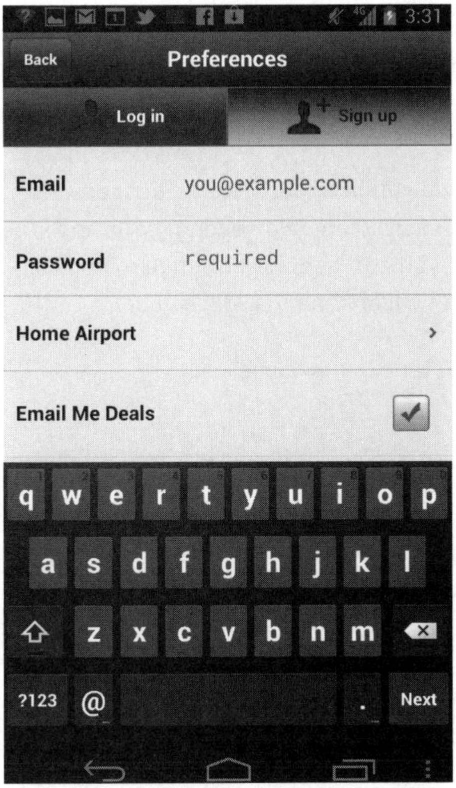

FIGURE 10.43: The Kayak app uses explicit input masks.

This explicit input mask reduces input errors because it shows the field format ahead of time and also keeps the label present during data entry. Of course, nothing comes free, particularly on mobile devices (just ask AT&T about their texting plans), and the resulting form sacrifices some of the length of the label. And some minimalists would argue this format also adds extra "noise." I disagree. Although the Kayak form does looks a bit busier than the native Android form designs, it is a small price to pay for improved clarity.

When and Where to Use It

Any time you need your customers to enter data structured in some way, you should ideally show the desired format of the field inside the field as an Input Mask pattern and launch the appropriate soft keyboard.

Why Use It

Mobile devices are notoriously poor at capturing keyboard input. Fat fingers, multitasking, and being jostled in the bus or metro are all contributing factors to this issue. Input masks reduce data entry errors and therefore should be used whenever possible to produce a better mobile experience.

Other Uses

As discussed earlier, for fields that allow a wide range of possible values, an implicit simple mask such as Phone Number is sufficient. However, if your app requires enforcing a specific format for entry, you can use the effective experimental variation, as shown in a high-resolution wireframe in Figure 10.44.

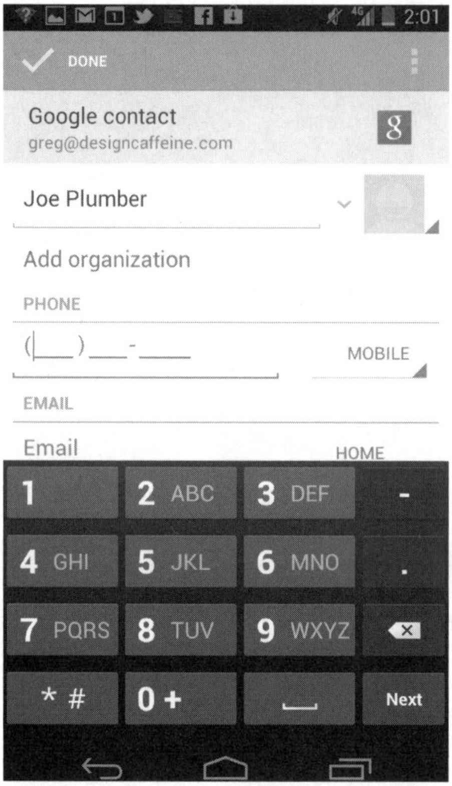

FIGURE 10.44: This experimental pattern puts a static input mask inside the Android 4.0-Style phone number field.

Compare this wireframe with Figure 10.42. Interestingly, this format actually reduces clutter because the input field no longer carries the same label (Phone) twice: once inside the field and once above the field. Luke Wroblewski first proposed this idea in 2010 at his Mobile Input workshop at Design4Mobile in Chicago (http://static.lukew.com/MobileInput_LukeW.pdf). This is an excellent approach to implement the Text Input with Input Mask pattern because it reveals the format of the expected entry ahead of time, so there are never any surprises about the strange separator characters popping up seemingly at random all over the place (as discussed later in the "Caution" section). Unfortunately, to this day, this remains largely an experimental pattern and has not gained wider adoption.

Pet Shop Application

For the Pet Shop app, you can use a different tactic for the Phone Number field in the Pet Registration Form. Because the app accepts both U.S. and international phone number formats, you can choose to make the input mask simply say 111-111-1111 (see Figure 10.45). This should accommodate both the formats with and without the country code and phone number extensions.

On the other hand, the second field is the Social Security number of the pet's owner: a well-formed nine-digit number. For this field use Luke Wroblewski's idea of a static input mask inside the field.

FIGURE 10.45: This wireframe for the Pet Shop app demonstrates implied and static input mask patterns.

Tablet Apps

The input mask applies even more strongly to tablets because the screen has plenty of real estate to provide a separate label and a static in-label input mask. Unfortunately, most native reference form implementations (such as Calendar) fail to take advantage of the cornucopia of screen space and consequently do not provide additional input instructions, even when abundant space is available. Instead, the Contacts app in the Galaxy Tab 2 7-inch Tablet with Android 4.0 reveals the format gradually using a loose dynamic input mask. As shown in Figure 10.46, the system initially removes a perfectly good "–" separator character that the person initially types as part of 415–222 and replaces it with the ( ) format as the customer continues to type (415) 222–33, surprising him with the additional characters popping up in the field. On the other hand, as shown on the right side of the figure, the Galaxy Tab 2 tablet also seems to enable all sorts of nonsensical user formatting that can be used to, seemingly at random, override the native dynamic input mask.

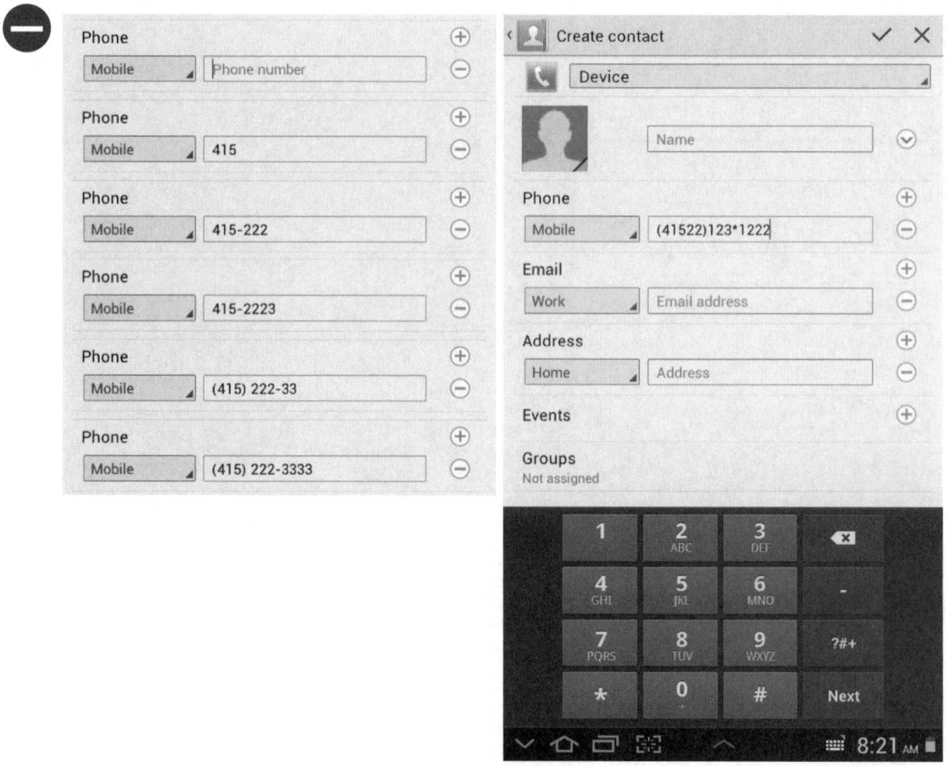

FIGURE 10.46: This antipattern in the native Calendar app on the Galaxy Tab 2 uses a loose dynamic input mask.

Sometimes the system overrides the format the person puts in, and sometimes it doesn't, creating a confusing experience. This implementation is not recommended. It is always best to reveal the formatting information upfront, if possible.

! Caution

Although this pattern is thankfully becoming rare, some apps *strictly* auto-format the text input as it is typed; that means the system immediately removes any characters it deems inappropriate, *even when they are part of the mask*. As Luke Wroblewski so eloquently showed in his article "Input Masks Design," (December 13, 2008, http://www.lukew.com/ff/entry.asp?756), this kind of strict dynamic input mask is actually an antipattern because it creates a confusing interaction when handling in-field characters that are part of the mask.

For example, as shown in Figure 10.47, while entering $60,000.00 into a strict dynamic input mask field, the system swallows the $ character after it's typed and further formats the commas in a way that's entirely different from the decimal point, which all adds up to a confusing interaction.

US Dollar:

$

Number: $#,###.00

US Dollar:

Number: $#,###.00

US Dollar:

$6

Number: $#,###.00

US Dollar:

$600

Number: $#,###.00

US Dollar:

$6,000

Number: $#,###.00

US Dollar:

$60,000

Number: $#,###.00

FIGURE 10.47: It's an antipattern for a strict dynamic input mask to remove characters as they are typed.

Another thing to keep in mind is to not go overboard on the keyboard specificity. For example, the specialized time entry keyboard from the Pocket Informant Android App, as shown in Figure 10.48, actually tries to limit the hour entry to the numbers 1 and 0 for the first digit then to the numbers 0 through 9 for the second digit, which depends on the first digit typed. This is presumably meant to help people avoid typing nonsense entries such as 91 hours. Unfortunately, the result is rather confusing because keys on the keyboard light up and go dim for no apparent reason while the customer enters even a simple time such as **8:00 AM**, which distracts the customer greatly from the primary task.

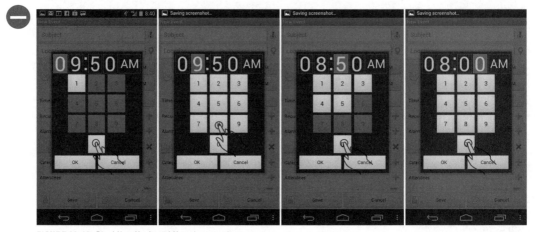

FIGURE 10.48: Disabling Keyboard Keys is an antipattern.

Avoid disabling the specific keyboard keys because it is overkill and a (fortunately rather rare) antipattern.

Finally, at all costs avoid breaking up fields into their component parts. For example, don't abstract the U.S. three-digit area code into a separate field in front of the phone number. This is an antipattern and creates a confusing interaction. Fortunately, this has become rarer in recent years.

Related Patterns

10.9 Pattern: Textbox with Atomic Entities

10.7 Pattern: Free-Form Text Input and Extract

10.9 Pattern: Textbox with Atomic Entities

Some textboxes are used as a kind of search box to locate and enter specific discrete, nondivisible system objects, also called *atomic entities*.

How It Works

The customer starts by typing a few characters that identify the atomic (also called *discrete*) entity, and the system performs a dynamic look up, returning the auto-suggestions, usually as a drop-down below the textbox. As soon as the customer taps one of the auto-suggestions or the typed characters resolve themselves into a single entity, that entity is added to the textbox. If multiple atomic entities are allowed in the same search box, the focus remains in the textbox, enabling the next text entry, which is input the same way.

Example

An excellent example of this pattern is the Invite field in the native Android Calendar app (see Figure 10.49). The customer can enter one or more people who are pulled in dynamically from the Contacts database.

If one of the contacts is not recognized (for example, Joe Plumber), the system forms a red box around the unrecognized contact. The customer can then edit the name or delete the entire entity with a single tap of the Delete button on the keyboard.

When and Where to Use It

Use this pattern any time you enter one or more discrete entities from a long collection (50 or more objects). Typical uses include Contacts, Social Media Connections, Countries, and Airports.

Why Use It

As the information in the world becomes richer and more ubiquitous, more collections begin to qualify as "large," so they become awkward to enter using a drop-down/select control. Textbox with Atomic Entities pattern provides a robust solution for this problem. You can use it to search the server or operate on local databases that are cached on the mobile device entirely on the client side.

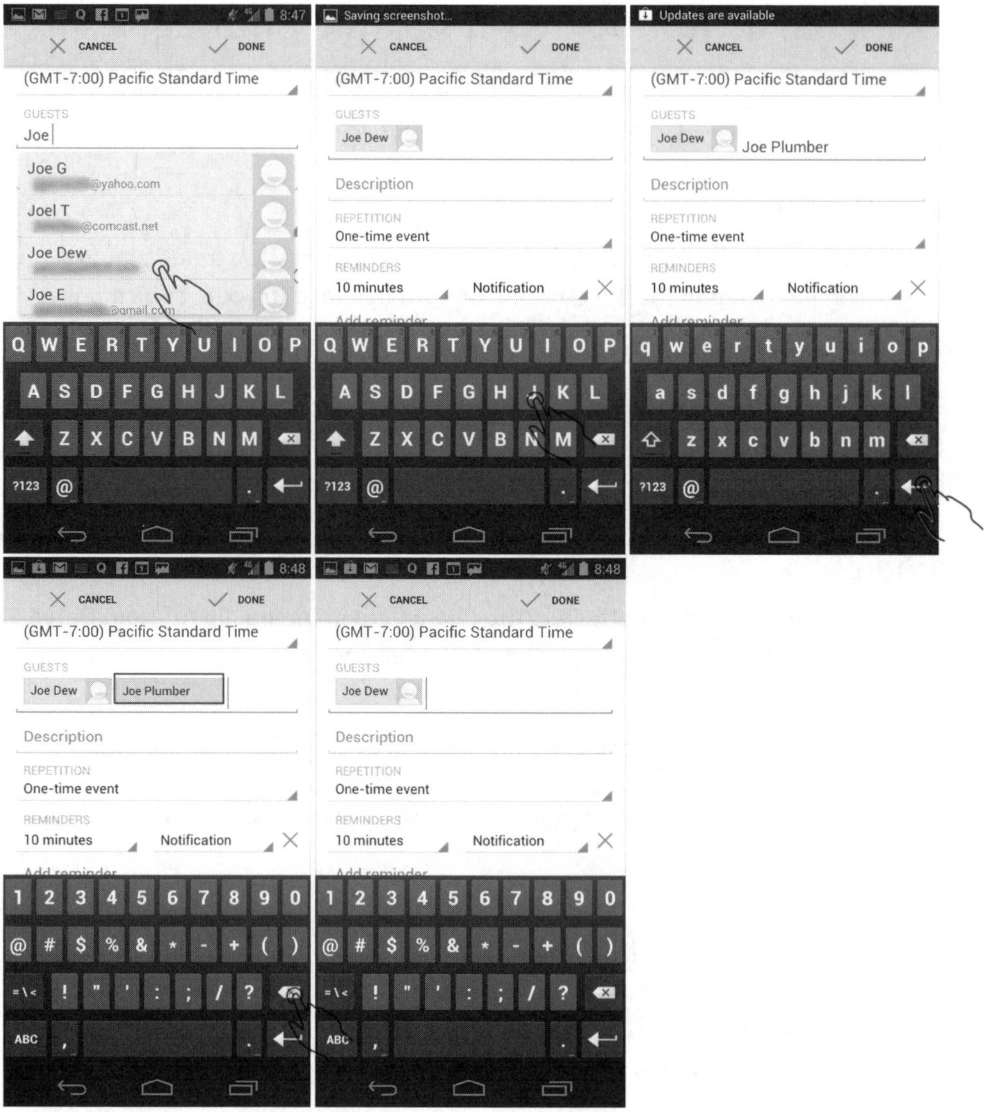

FIGURE 10.49: The textbox with Atomic Entities pattern works well for adding Contacts to a Calendar Invite.

Other Uses

Smashing Magazine published an article that provided an excellent argument for using an Atomic Entity approach for a control that can be used to enter one of today's 249 countries, rather than using an overly long drop-down ("Redesigning The Country Selector," by Christian Holst, November 10th, 2011, http://uxdesign .smashingmagazine.com/2011/11/10/redesigning-the-country-selector/). Incredibly, no one has yet implemented this Country Selector in the mobile app. The text field auto-suggest search Christian created is robust and intuitive, surfacing more common items (United States) higher in the list than the less common ones (United Emirates). Add to this a few techniques discussed in Chapter 9, "Avoiding Missing and Undesirable Results," such as a quick controlled vocabulary substitution (Holland to Netherlands), and you have a chance to create an industry-defining mobile experience. Other capabilities of this pattern include the ability to enter new entities on-the-fly, as dictated by the purpose of the app. For example, the Calendar app enables the entry of a well-formed e-mail address directly, instead of picking an existing entity from the Contacts database.

Pet Shop Application

According to the current version of Wikipedia, there are from 400 to 600 different dog breeds. When registering a dog in the Pet Shop app database, you can use the list to form a controlled vocabulary from which the owner can select a breed of a dog. The wireframe set is fairly straightforward (see Figure 10.50).

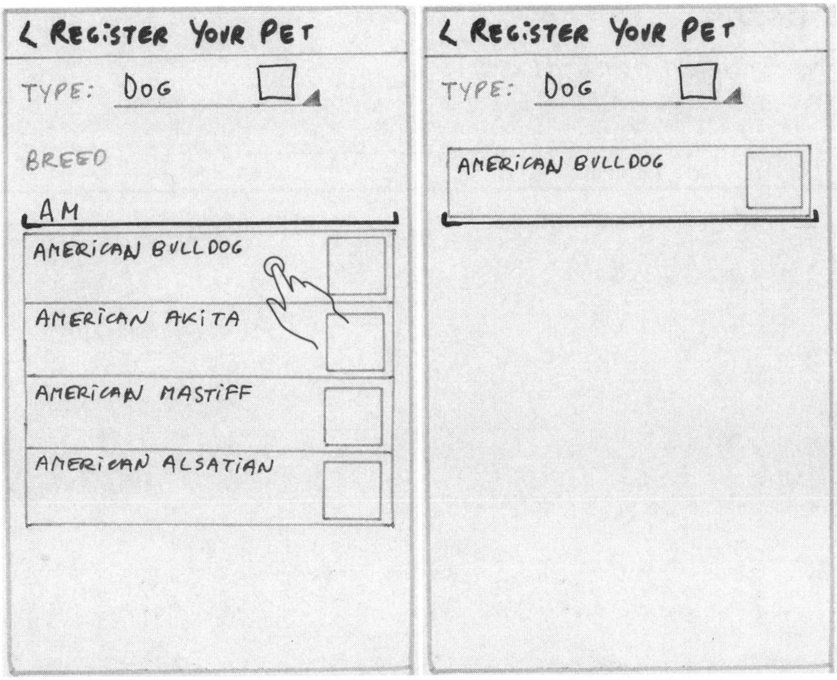

FIGURE 10.50: Adding a dog breed using the textbox with the Atomic Entity pattern is quick and intuitive.

The owner starts typing the first few letters, and a list is shown to him of the dog breeds that match the entry. The owner can then continue typing (up to the full name of the breed) or simply select the breed he wants from the list of auto-suggestions. The pet breeds do not change often, so these can be cached in a small database on the mobile device, providing a smooth and responsive experience.

Tablet Apps

This pattern should work the same for the tablets as it does for the mobile apps but fit more auto-suggestions. Remember that depending on the orientation, the auto-suggestion layer can display below the field (vertical tablet orientation) or to the right of the field (horizontal orientation). Don't get caught up with following the mobile implementation.

! Caution

Remember to test boundary conditions. What happens when the invite has more people than can fit on a single line?

Related Patterns

10.7 Pattern: Free-Form Text Input and Extract

CHAPTER 11

Forms

Data entry on a small screen is fascinating, but customer inputs must be combined together in forms to ultimately enable interaction. Luke Wroblewski made an eloquent statement about filling out forms in *Web Form Design: Filling in the Blanks* (2008, Rosenfeld Media): "Forms suck." That's the reality. This chapter is mostly about how to make them *suck less* on Android devices—and in some cases, even make filling out Android forms downright fun.

11.1 Pattern: Inline Error Message

Whenever an error occurs when you fill out a form, the Inline Error Message pattern provides the standard way to help the customer out of it.

How It Works

When an input error occurs, the system notifies the customer which fields need to be corrected. Generally you can recognize two components of the Inline Error Message: a field-level error indicator and a general error message (usually on top of the screen).

Example

Visual error indicators differ greatly between apps. My favorites are positive visual indicators such as icons combined with the traditional red color, such as the reference implementation from the registration flow of the Calendar app (see Figure 11.1), which adds a red round (!) icon on the right side of the textbox if the person attempts to move past the e-mail field.

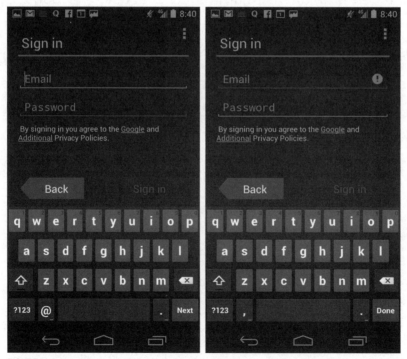

FIGURE 11.1: This implementation of the Inline Error Message pattern is from the Calendar app.

The Calendar app does not show a message in addition to indicating a missing value in the field. Although it's often preferable to show the message and indicate which field has an issue, sometimes the answer is obvious, especially for shorter forms.

An alternative to this example is the eBay app's registration form. In Figure 11.2 you can see that the form messages field-level errors by putting a red box around each field and places a composite error message on top of the form in red font.

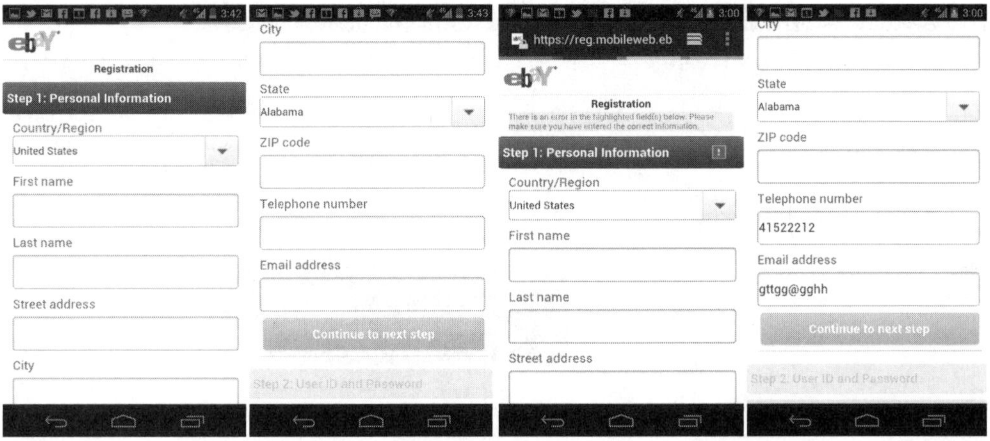

FIGURE 11.2: The Inline Error Message pattern in the eBay app uses two ways to alert the customer of issues.

In the eBay implementation of this pattern, *all* the fields that have an issue receive the red border. This is the correct way to implement this pattern.

Both the Calendar and eBay forms disable the Next/Continue button, which provides a further indication that something has gone awry and the form is not yet ready for submission. Read more about this in the "11.5 Pattern: Cancel/OK" section.

When and Where to Use It

Any time you need to show your customers that they made an error in the form, this is the default pattern to use.

Why Use It

This pattern has stood the test of time and has proven to be the most usable way to show form errors.

Other Uses

One potential modification that you can make for long forms, such as the eBay registration, is to show the error message as a toast alert (see the next section, "11.2 Pattern: Toast Alert"). The advantage of doing this is that the error message is shown to customers even though they have scrolled down past the message (which is hidden from view on top of the page). Compare the toast alert wireframe shown in Figure 11.3 to the message on top of the page shown in Figure 11.2.

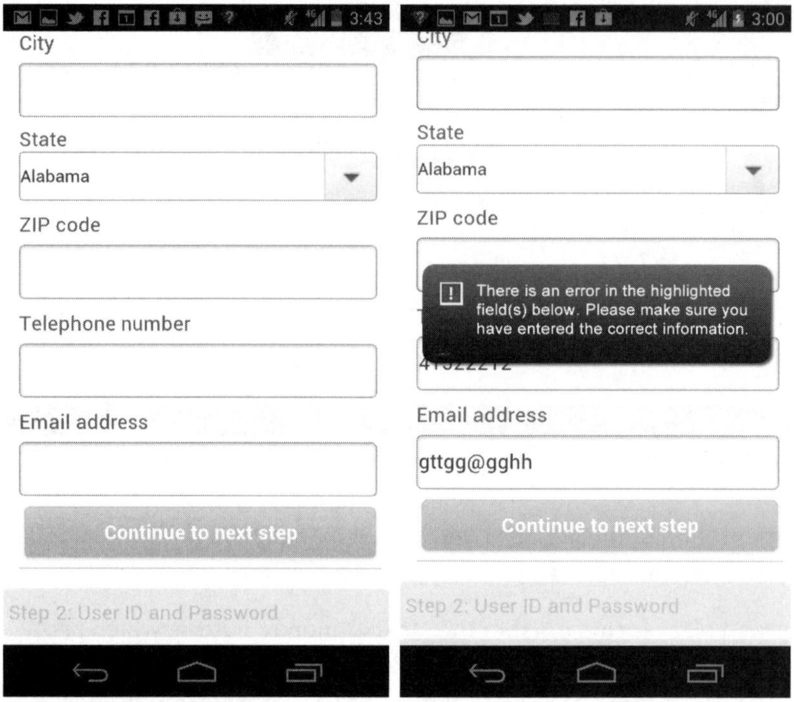

FIGURE 11.3: This proposed wireframe provides an alternative implementation (the Toast Alert) of the eBay app's Inline Error Message.

One disadvantage to using the Toast Alert pattern with the inline error message is that the overall error message is not "resident." In other words, the alert disappears after a few seconds, which could be confusing to customers if the message is long and contains detailed instructions as to how to solve the issue. For example, if your password error message reads something like "Your password needs to be 8 characters long and contain 3 special characters, 3 numbers, and 2 letters, upper and lowercase, and it cannot be the same as 10 previous passwords, nor can it contain the name of a movie character like C3PO or Darth Vader or Han Solo or Leah Buns or...," you may not want to use the Toast Alert. (And you may just want to consider making it a wee bit shorter and easier for your customers. See the "11.2 Pattern: Toast Alert" section for more discussion.)

Pet Shop Application

Because I am a fan of short registration forms and error messages on top of the page, the example shows you how to sketch a simple inline error with the top message for registering with the Pet Shop app (see Figure 11.4).

The form takes advantage of having an e-mail address on hand on the mobile device, so it prefills the e-mail address field but still enables the customer to change it if necessary. The app uses phone device ownership as a two-factor authentication pattern (read more in Chapter 12, "Mobile Banking") and use the Login Accelerator pattern (see the "12.1 Pattern: Login Accelerator" section in Chapter 12) to secure the account with only a four-digit combination.

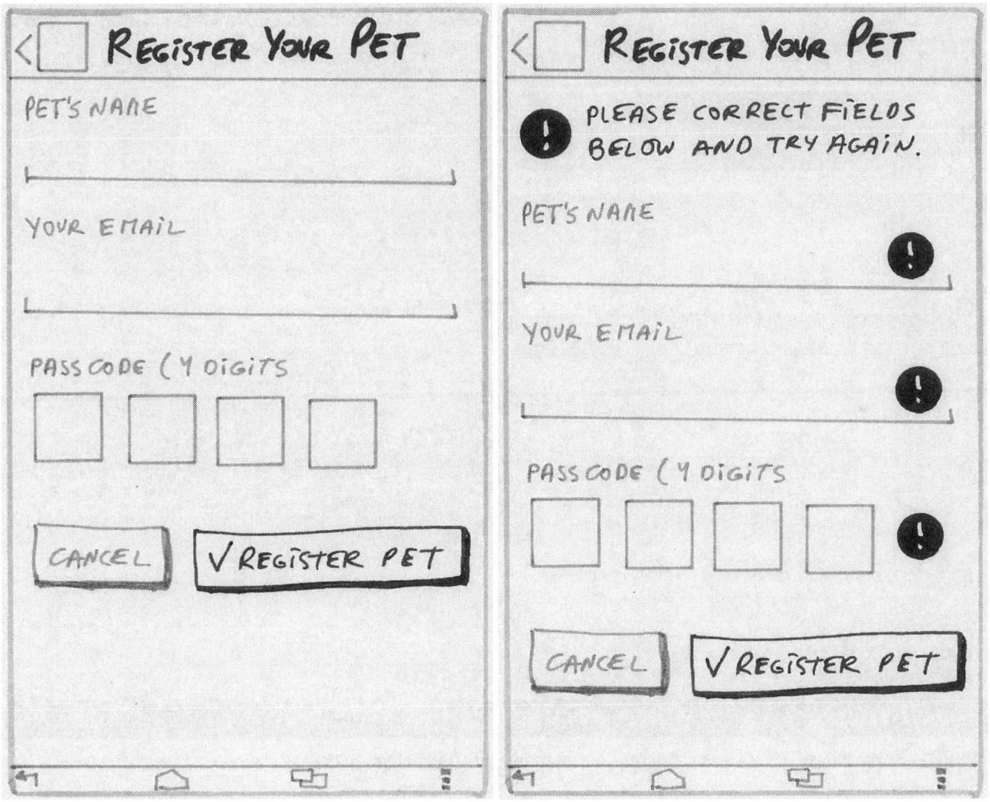

FIGURE 11.4: This wireframe of the Inline Error Message pattern is for the Pet Shop app user registration form.

In the left wireframe, the customer just started to fill out the form. On the right side you see the form with the inline error, as the customer missed entering the name of his pet, his e-mail address, and the four-digit PIN, all of which are required. I used the same round red icon on the top level message to indicate the field-level error. This is intentional—both icons are similar in shape and color to draw attention to the tie-in of the message above the form to all the error fields below.

Tablet Apps

In tablet apps, there is a bit more room, both vertically and horizontally. Because of that, the message on top of the page is easier to find (see Figure 11.5) and is most likely to be displayed above the fold, making the toast alert modification described earlier unnecessary.

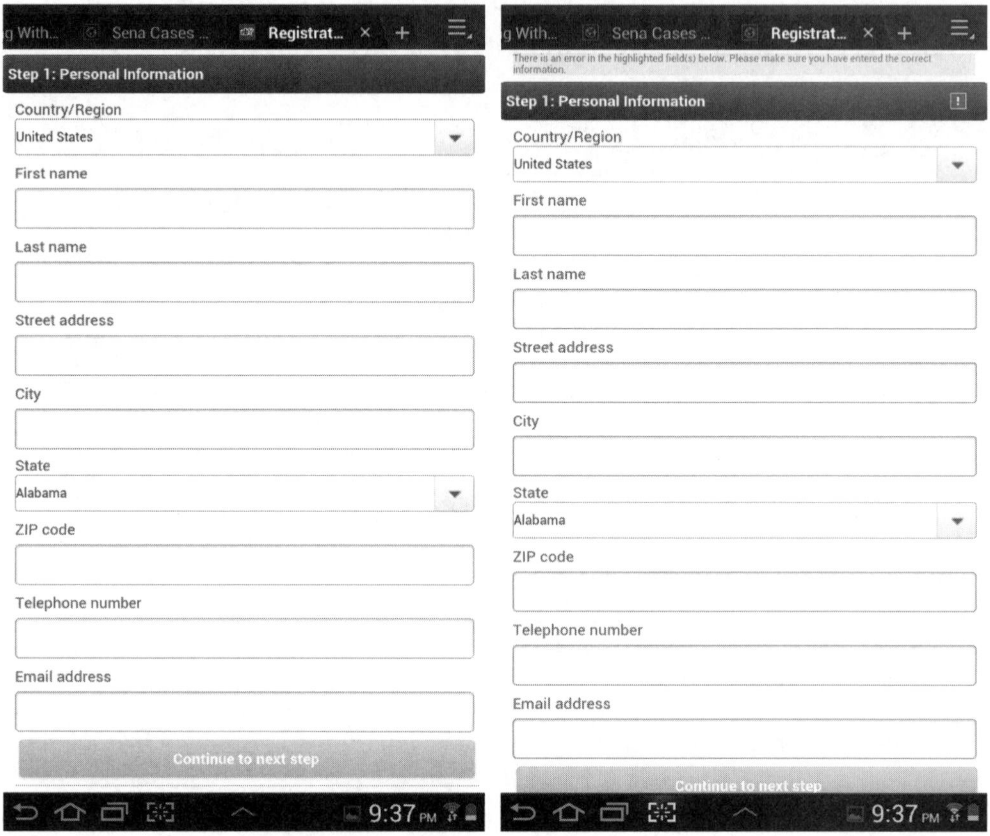

FIGURE 11.5: The Inline Error Message pattern in the eBay app registration form works even better on a 7-inch Galaxy Tab 2 with Android 4.0.

Another thing all that wonderful extra space lends itself well to is more detailed instructions located next to the field. Depending on the device orientation and form design, these can be shown on the right of the label or immediately below it. Figure 11.6 shows a wireframe of the example of how that might look in a vertical (top) and horizontal (bottom) tablet orientation. Unfortunately, today this largely remains an experimental pattern.

ZIP code

Like 11111 or 11111-1111

ZIP code

Like 11111 or
11111-1111

FIGURE 11.6: These proposed wireframes of the Inline Error Message pattern in the eBay app registration form provide extra field-level help information.

! Caution

Beware using only color gradations to indicate problem fields, as eBay has done. People that are color-blind (which is approximately 3 percent to 5 percent of the population, depending on gender, location, and other factors) may have trouble picking out the fields with a red box around them. A good test is to print the error and nonerror states in black and white to check if there is enough contrast between the states. As you can see with the eBay form example, there is enough contrast between the error and nonerror states (see Figure 11.2), though just barely. When in doubt, it is best to use a secondary positive error indicator, such as an icon in the field or next to it, as shown in Figure 11.1 and in the Pet Shop app wireframe in Figure 11.4.

Related Patterns

11.2 Pattern: Toast Alert.

11.2 Pattern: Toast Alert

Toast Alert is named thus because in the original implementation the alert layer would pop up from the bottom of the screen like a piece of bread popping out of the toaster. The actual transition of the alert popping up is much slower than the bread ejected from your toaster—the latter always seems to fly out with the force equal to that of a conductive payload coming out of the railgun, but hey, you can't beat those sweet Android food metaphors. The current Android 4.0 implementations of the Toast Alert have a variety of entrance motion variations, including just appearing on the screen. Today, the name "Toast Alert" is used mainly as a way to distinguish general messages that do not have buttons and go away by themselves after a short time from so-called Pop-up Alerts (see the next pattern) that do have buttons and remain on the screen until customers take action to dismiss them.

How It Works

When the condition that generates the toast alert is met, the screen displays a small overlay window with the specific message, icon, or both. After a specified time period (usually a few seconds) the alert window disappears. The alert window can also be dismissed before the time period has elapsed by tapping on the alert window or anywhere else on the screen. The entrance and exit of the alert window are sometimes performed as an animated transition, which may include darkening of the background application window.

Example

One example of this pattern is a white Toast Alert that comes up when the Trulia app loses a network signal (see the left side of Figure 11.7). Its counterpart on the Kayak app is black (which is pictured on the right side of Figure 11.7).

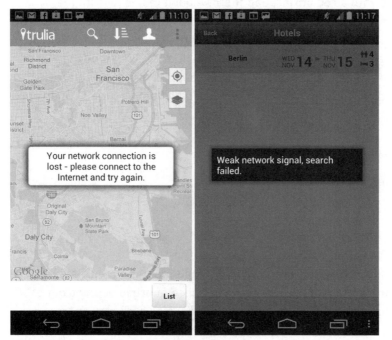

FIGURE 11.7: The Trulia (left) and Kayak (right) apps use the Toast Alert to indicate a weak network signal.

The more "traditional" implementation of the Toast Alert used for confirmation comes from the LinkedIn app. The Alert comes up if the customer has any messages and after accepting an invitation to connect, for example. In both cases, the alert comes up from the bottom and overlays the results. (See Figure 11.8.)

The Toast Alert pattern is great to communicate confirmation of a dynamic system action that requires a network call.

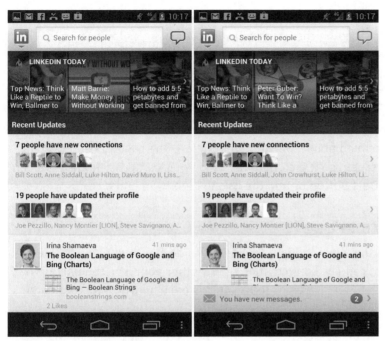

FIGURE 11.8: The "Traditional" Toast Alert comes up on the bottom of the page in the LinkedIn app.

Yet another implementation of this pattern comes from the Amazon Fresh app (see Figure 11.9) when the customer adds an item to the cart. Note the Toast Alert actually comes up on top of the page, pushing down the search results, so the toast falls out of the toaster, as it were.

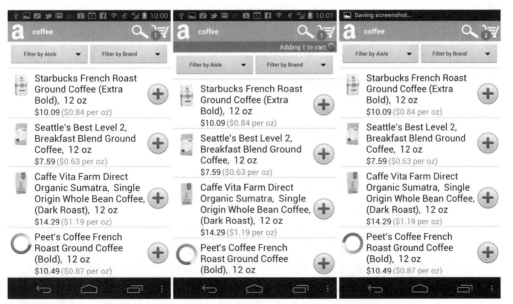

FIGURE 11.9: The Amazon Fresh app's confirmation Toast Alert displays on the top of the page.

Because the alert is on the top of the page and (though you can't tell) it's actually a solid red high-contrast color, the fingers never cover it up, as they do on the LinkedIn app. This enables Amazon Fresh to keep the alert small, so it does not cover up any results, and the customer can keep on shopping without any interruptions. The alert is also close to the shopping cart icon, the badge on which gets incremented each time the alert pops down, reinforcing in the customer's mind the tight relationship between the alert and the count of items in the shopping cart. There is one bad thing about this implementation: The results keep moving up and down on the screen; so if the customer is too quick with her finger, she can miss the next Add to Cart button or add the wrong item by mistake.

When and Where to Use It

Any time you have a transient condition that needs to be communicated to the customer but does not require confirmation or any other action on the part of the customer, use the Toast Alert pattern. Toast Alert is particularly good as a confirmation device.

Why Use It

The Toast Alert pattern juxtaposes nicely against its ugly stepsister, the Pop-up Alert. Unlike the Pop-up Alert, the Toast Alert doesn't need an explicit function to be dismissed; it just goes away by itself, which is convenient on a small mobile screen.

Other Uses

See "11.4 Pattern: Callback Validation."

Pet Shop Application

Figure 11.10 shows the Toast Alert that comes up after the item has been added to the shopping cart using the interface shown. The Toast Alert comes down from the top of the page (refer to the Amazon Fresh example in Figure 11.9) but this time, the alert comes *on top* of the search results, partially covering up the top result on the page (instead of moving the results down, as Amazon Fresh does).

Thus you still get the benefits of the alert that comes close to the shopping cart icon and one that is not covered up by fingers holding the device. However, you also solve any issues that involve wrong buttons being pressed due to the search results inventory moving up and down. The only thing that could be the fly in the ointment in this case would be that the top search result is partially covered up. However, due to the nature of the Toast Alert, it is easy to dismiss it by tapping anywhere on the screen, so if the customer wants to get to that first result, all she has to do is tap anywhere onscreen (including the alert), or simply wait 1 second, and the alert disappears, so the result becomes accessible again. The message can be varied so that it does not get boring.

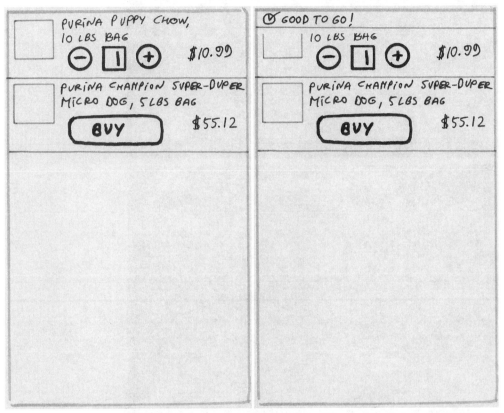

FIGURE 11.10: The Alternative Confirmation Toast Alert drops down over the top of the search results in the Pet Shop app.

Tablet Apps

In tablets even more than on mobile devices, you must be concerned about fingers covering up the alert. You also have another problem to deal with: A much larger screen, which means the customer's attention can be anywhere, and she may not see the alert. That said, an alert that comes down from the top middle of the screen, regardless of the orientation, is usually a safe way to go, especially if it's done in contrast colors and includes a quick, smooth animated transition. Figure 11.11 shows one example from the N.O.V.A. game on the tablet.

FIGURE 11.11: Here is the Welcome Toast Alert on the top of the page in the N.O.V.A. Game app on a tablet.

! Caution

A cautionary example of using this pattern is as a warning that additional actions are needed, as exemplified by its use in the Peapod app (see Figure 11.12) to indicate that the quantity is too high and the customer needs to call customer care. This use of the Toast Alert is an antipattern because this alert text is complex, yet by its nature, the Toast Alert disappears too quickly to read and fully grok the message properly. In addition, Toast Alert is particularly poor for recommending further actions because traditionally tapping the alert or any negative space around it should simply dismiss the alert, leaving the confused shopper to wonder where exactly does she need to go from here to call customer care.

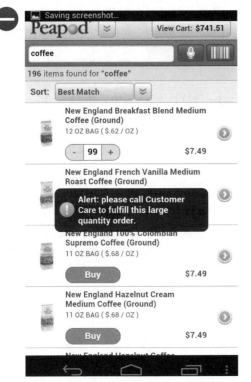

FIGURE 11.12: This long and complicated Peapod app Toast Alert that requires customer action is an antipattern.

If you need to give your customer a choice of whether to perform a certain action, a Pop-up Alert, which is discussed in the next section, makes a better choice.

Related Patterns

11.3 Pattern: Pop-up Alert

11.3 Pattern: Pop-up Alert

Earlier in the chapter, this pattern was called "the ugly stepsister" of the mobile alert magic universe. This is because everyone and their evil stepmother try to make a Pop-up Alert into a princess, which it isn't. However, used properly, Pop-up Alert reveals its inner beauty as the industrial-strength flow disrupter.

How It Works

When an alert condition occurs, the system pops up a lightbox, while also often darkening the background of the current task. The popup requires the customer to choose one of the actions that displays on the alert, which use up to three buttons on the bottom of the popup window.

Example

A typical implementation of a three-button Pop-up Alert is shown in Figure 11.13.

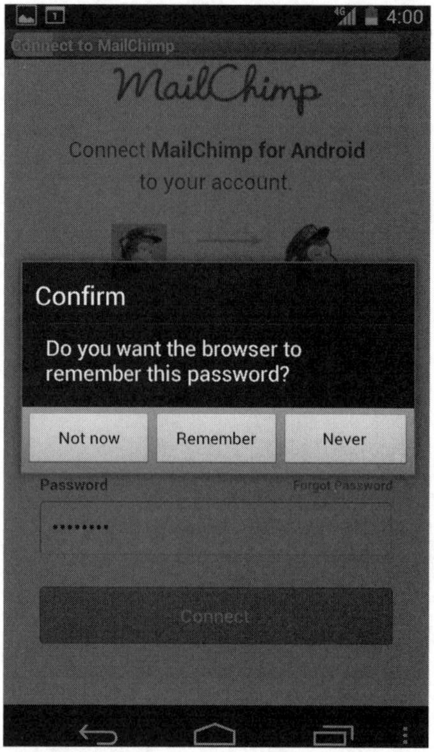

FIGURE 11.13: The Mailchimp app includes a typical Pop-up Alert with three buttons.

Note that the alert stops the proceedings, requiring the customer to take one of the three actions offered by the popup.

Another appropriate way to use the Pop-up Alert is as a warning about the system state; as shown in Figure 11.14, it warns about the battery running low.

In this case, the task is interrupted to convey the urgency of the message and to convey a call to external action: Connect charger.

FIGURE 11.14: This system Pop-up Alert warns about a low battery.

When and Where to Use It

Use this pattern when you need to disrupt the current task flow with an alert that requires the customer to take a (preferably urgent) action. Pop-up Alert is the atomic bomb of the alerts arsenal; use it with caution.

Why Use It

Sometimes things go wrong and you do need to shout "Stop the presses!" The Pop-up Alert does the trick and suspends all other operations of the device until the customer takes some action.

Other Uses

One of the often overlooked ways to use the Pop-up Alert is as a rather benign Welcome tutorial, such as the example in Figure 11.15 from the native Maps app.

Another way is to use it to display the legalese of the terms and conditions, as shown in Figure 11.16 for the Google Plus app.

FIGURE 11.15: You can use the Pop-up Alert as a Welcome tutorial this example from the Maps app.

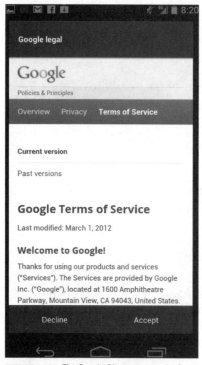

FIGURE 11.16: The Google Plus app uses the Pop-up Alert pattern for the terms and conditions.

For more information about terms and conditions and Welcome tutorials, see Chapter 5, "Welcome Experience."

Pet Shop Application

In the Pet Shop app, you can use the Pop-up Alert to give the customers a reminder of the benefits that come with registering their pet. (See Figure 11.17.)

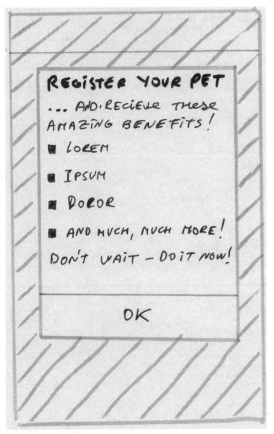

FIGURE 11.17: The Pop-up Alert is used as a reminder to register in the Pet Shop app.

The reminder has only a single button—OK—and is shown only one time when the app first loads.

Tablet Apps

Tablet app designers must be especially careful with the use of the Pop-up Alert because it can take over the entire large surface area of the tablet. (See Figure 11.18.)

The buttons on the bottom are fairly awkward to tap and can be hard to find. If you do end up using this pattern, consider smaller size modal windows that sit on top of the existing content background, leaving some empty space around the edge of the popup.

FIGURE 11.18: The terms and conditions in the eBay app on a 7-inch Galaxy Tab 2 display as a Pop-up Alert.

! Caution

In his seminal book *About Face* (2007, Wiley), Alan Cooper famously called the Pop-up Alert pattern "stopping the proceedings with idiocy." That's a perfect description of using the Pop-up Alert incorrectly, and there are many ways to do this, so much so that I selectively chose the examples for this "Caution" section.

One of the common antipatterns is discussed in Chapter 9, "Avoiding Missing and Undesirable Results." It uses the Pop-up Alert to show no results conditions, as the Booking.com app does in Figure 11.19.

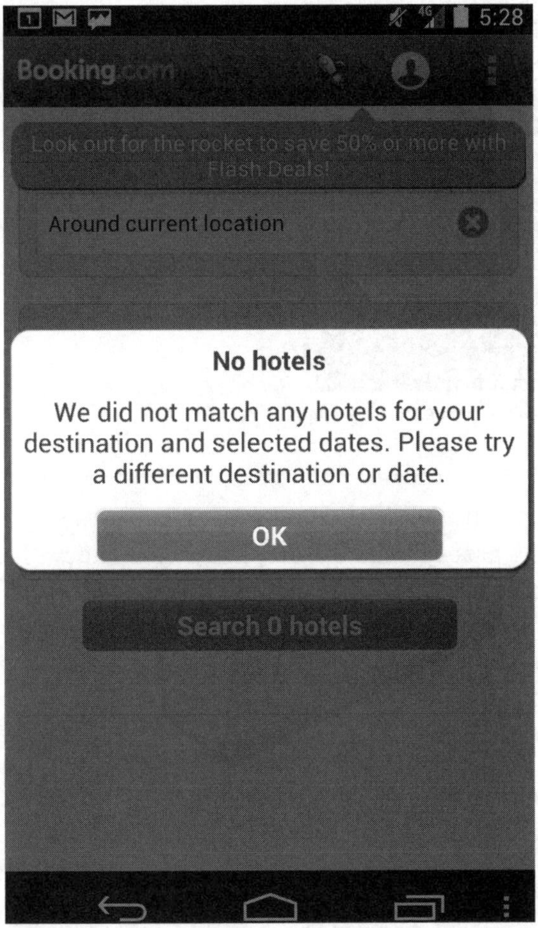

FIGURE 11.19: The Pop-up Alert can be an antipattern if it shows a no results condition as it does in the Booking.com app.

Another misuse of the Pop-up Alert is to use it to show that the wrong data has been entered via a wheel control, as shown in Figure 11.20. Refer to Chapter 10, "Data Entry," for the right way to show Date and Time wheel input issues.

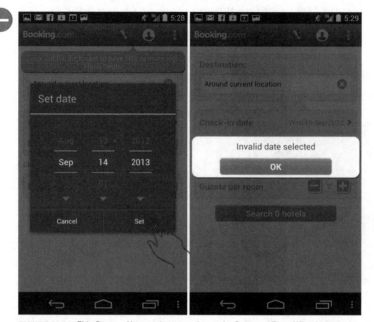

FIGURE 11.20: This Pop-up Alert antipattern shows the Date and Time Wheel input errors in the Booking.com app.

But the Pop-up Alert misuse prize goes to the Yelp sign-up form, which uses it to reveal form errors, one at a time. The proceedings are stopped by the idiocy of having a Facebook sign-up pop-up alert, which sets up the playing field nicely for all the other pop-up errors, that march across the mobile screen like an out of tune Highland Pipe Band. The Pop-up Alert Legion Antipattern is shown in Figure 11.21.

This is the equivalent of "slapping" your customer "with a splintered ruler," one strike for each mistake he makes (as Alanis Morissette so brilliantly sang in "All I Really Want"). This antipattern will annoy your customers and *greatly* negatively impact the completion rate of your form. As egregious as this sequence of Pop-up Alerts looks, this pattern is unfortunately fairly common. A much better alternative is to use the Inline Error Message pattern described earlier in this chapter; not only does it avoid the strike of the ruler (having to tap OK on the Pop-up Alert, which many people equate to acknowledging that they are "idiots"), but it also shows all the missing and error fields at the same time, so the customer can correct them all at once and resubmit the form correctly.

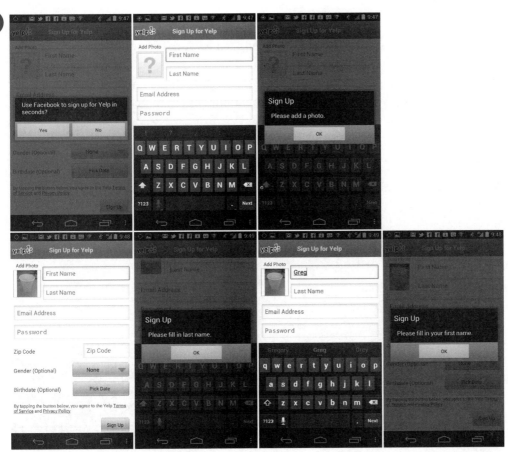

FIGURE 11.21: In the Pop-up Alert Legion antipattern, multiple Pop-up Alerts show form input errors one at a time in the Yelp app.

Related Patterns

11.1 Pattern: Inline Error Message

11.4 Pattern: Callback Validation

Sometimes, it's impossible to validate form inputs in-situ using only client-side code. The Callback Validation pattern is the mobile equivalent of the Ajax call that allows an asynchronous server trip for more sophisticated dynamic database-driven validation.

How It Works

When a chunk of customer-input data needs to be validated on the server, the system detects when the input is completed and issues an asynchronous server call to validate the data, returning with one of two states, OK or Fail. Fail states are often accompanied with the dynamic suggestions meant to help customers correct the error condition.

Example

A great example of the Callback Validation pattern is the Twitter app registration form.

FIGURE 11.22: The Twitter app includes an excellent implementation of the Callback Validation pattern.

The Twitter app waits for a delay of approximately ½ to ¾ of a second (500 to 750 milliseconds) in the typing of an *@username* and uses that delay as a starting point to issue the server-side lookup call to determine if the username is available. The typing delay is a more robust and satisfying way to issue the lookup call than waiting for the press of the Next button or an OnBlur event, even though it results in a lot of "false negatives" when the customer has not yet finished typing the full username.

When and Where to Use It

Any time the client-side validation does not provide enough information or robustness and a server-side trip is necessary, use the Callback Validation pattern as a preferred method of validation.

Why Use It

As the speed and availability of mobile networks keeps improving, the customers are beginning to have expectations of near-instant contextual feedback. The Callback Validation pattern satisfies their craving for nearly real-time system response by looking at the break in the typing pattern instead of waiting for the submission of the entire form to report the error.

Other Uses

Callback Validation is not just for usernames. You can use it for any kind of entry that needs to be validated using a server-side call—for example, airport names; hotel, flight, or car availability dates. Or you can use it for potentially any dynamic entity that can also be entered using a server-side variation of the Text Box with Atomic Entities pattern (discussed in Chapter 10).

Pet Shop Application

With more than 7 billion people in the world today, picking a clever yet unique username is becoming more and more of a challenge. Actually, it is a task that is important enough to use a dedicated page for picking a username and providing some auto-suggestions, as the hand-drawn example in Figure 11.23 demonstrates.

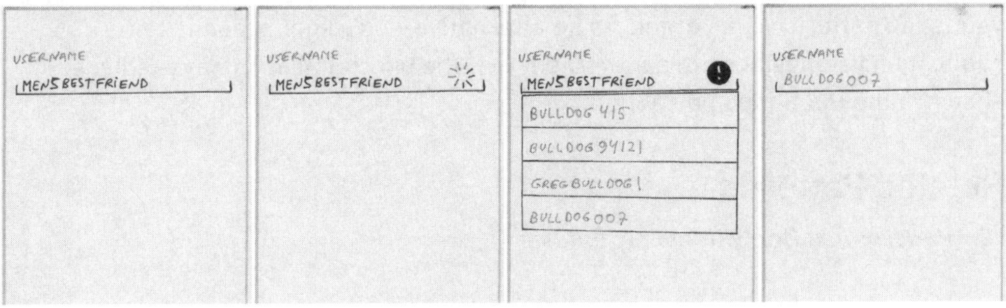

FIGURE 11.23: The Pet Shop app sports Callback Validation with auto-suggestions.

Just as the Twitter app detects a brief break in the typing, the Pet Shop app would do the same. However, in addition to saying that the username is taken, the app also offers some clever suggestions based upon the customer information entered in the previous screen (not shown). Data such as fragments of the area code, street address, ZIP code, pet type (dog, cat, or "bulldog") can all be used to help people create suggestions.

Tablet Apps

If you need your app to display auto-suggestions in addition to the dynamic validation, it helps to have a dedicated page for this type of interaction, as shown in Figure 11.23. However, a separate page is not necessary, especially on larger devices like tablets. Tablets should use this Callback Validation pattern more, which will probably occur in the near future. This pattern is especially pertinent for tablets because many of these devices operate on a strong Wi-Fi signal, rendering response capabilities that are similar to those of desktop web apps, where the Callback Validation pattern using Ajax is an accepted and expected part of the web.

❗ Caution

As with any mobile pattern that requires a server-side call, be aware that the signal may not go through, so build a robust backup strategy for validation using a server-side call after the entire form has been completed. It might show exactly the same UI as it would when the asynchronous call has been issued, even though the original asynchronous server call did not go through.

The Callback Validation pattern as used by Twitter strongly drives the "append" strategy of error recovery; in other words, the customer is most likely to recover from the error by appending additional characters to his username, as shown in Figure 11.22. If that is acceptable for your app, this version of the Callback Validation pattern works great. As an alternative, you might present some auto-suggestions alongside with a statement that the username is not available, as shown in the Pet Shop app section.

Related Patterns

10.9 Pattern: Text Box with Atomic Entities

11.5 Pattern: Cancel/OK

For designing forms, at some point in the process, clients inevitably ask, "Should I position the buttons as OK/Cancel or Cancel/OK?" This pattern describes how to position action buttons and hopefully to help you avoid 3-hour meetings devoted solely to this topic.

How It Works

Action buttons are positioned as Cancel/OK on the top or bottom of the form. The primary button is on the right and is sometimes larger or implemented with a more saturated color and/or icon. Often, the primary action button is disabled at the beginning of the process before a valid value is entered into the text field. (The last option is not recommended; see the "Caution" section.)

Example

The reference example of this pattern shows the action buttons positioned on top of the form in Android Calendar, as shown in Figure 11.24. The buttons remain on the screen even when the form scrolls up and down.

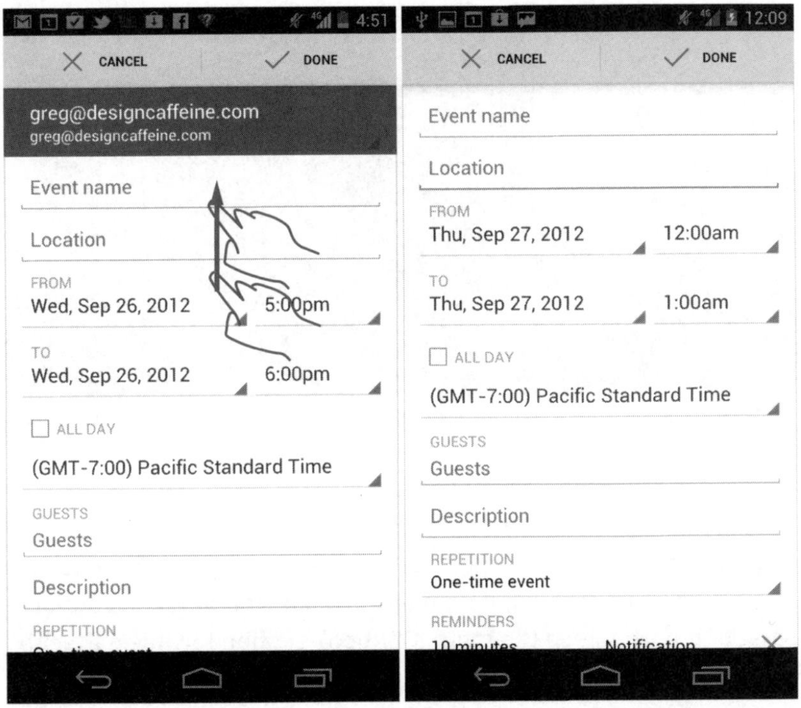

FIGURE 11.24: This reference implementation of the Cancel/OK pattern is from the Calendar app.

Even though both buttons have the same gray color saturation, they have different icons, with the OK (Done) presented as the check mark and Cancel as an X.

The reference implementation of lightboxes also follows the Cancel/OK pattern, as shown in Figure 11.25 for the Calendar and Contacts apps.

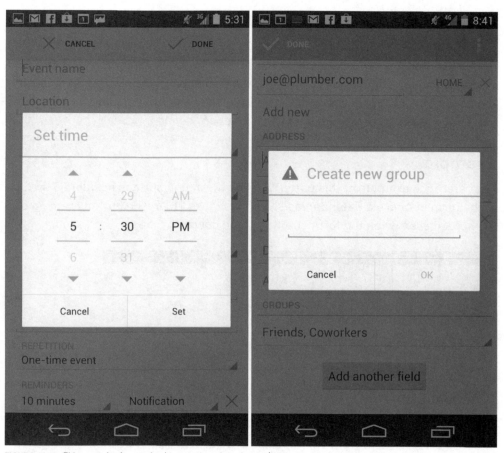

FIGURE 11.25: This second reference implementation of the Cancel/OK pattern is shown in lightboxes from the Calendar (left) and Contacts (right) apps.

The OK button is disabled for the Contacts app lightbox because there isn't yet a valid value in the form field. This is a common part of the pattern for simple forms or those that implement the Callback Validation pattern, but you must be cautious when disabling the action button (see the later "Caution" section for this pattern).

An alternative implementation with the buttons on the bottom of the screen is the search form from the Trulia app, as shown in Figure 11.26.

Even on the black-and-white picture you can tell that the primary action button (Find Homes) is much wider and more saturated in color (in real life it's dark orange) than the secondary button. Also Trulia uses descriptive verbiage on the button instead of the generic Search or OK.

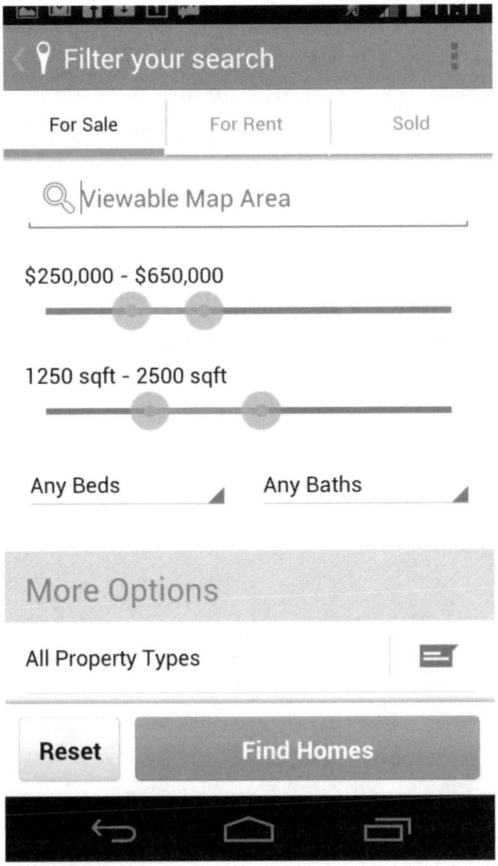

FIGURE 11.26: This form (with buttons at the bottom) from the Trulia app is a great alternative Cancel/OK implementation.

When and Where to Use It

This is the standard pattern that should be used for every form in your application.

Why Use It

Why implement action buttons as Cancel/OK instead of vice versa? This is not an idle question. Ergonomically speaking (read more about device ergonomics in Chapter 3, "Android Fragmentation"), the button on the left is easier to tap when operating the mobile device one-handed. Why would you want to override this natural convention and switch the buttons? The Cancel/OK convention on mobile stems from the early mobile phone app designs, which come from the western convention of reading from left to right. Following this convention, buttons on the left (Cancel) usually take you closer to the top menu or home, whereas the buttons on the right (OK) take you deeper into the Information Architecture (IA) of the device down to the leaf node of whatever function you are looking for. Thus the convention of Cancel/OK has taken a firm root in the smartphone design, even though some apps still refuse to follow it.

Other Uses

What should you do when you have only one button? If you follow the same convention (that the right side is equivalent to more or deeper into the IA), you should put it on the *right* side of the screen. However, for a single button, the ergonomic considerations often win, as in the example from the Contacts app shown in Figure 11.27, which places the button on the top *left*.

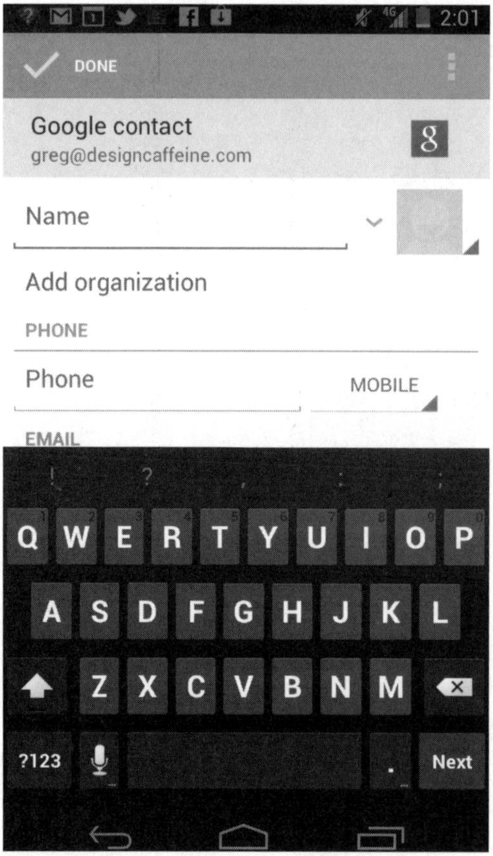

FIGURE 11.27: A sole Done button is on the left side of the screen in the Contacts app.

In the Android OS, the menu is traditionally assigned to the right side of the screen. Thus the menu button placement often wins the right side when competing with a sole action button (which then goes on the left).

Another great solution to the sole action button placement is to make a single large button across the entire width of the screen, as shown in Figure 11.28 on the Kayak search form.

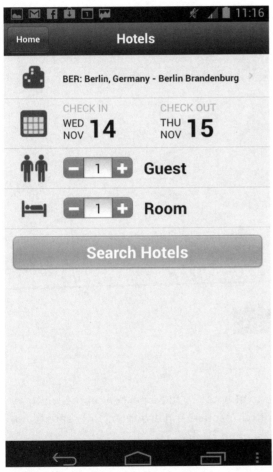

FIGURE 11.28: A sole OK button spans the width of the screen on the Kayak search form.

Unfortunately, although they're functional, large buttons that span the width of the device tend to look a bit goofy and old-fashioned, especially on larger tablet-like devices such as the Galaxy Note. If large cartoony buttons offend your sensibilities, you may want to consider making a slightly smaller button that can still be centered in the middle, as on the Contacts app (see Figure 11.29).

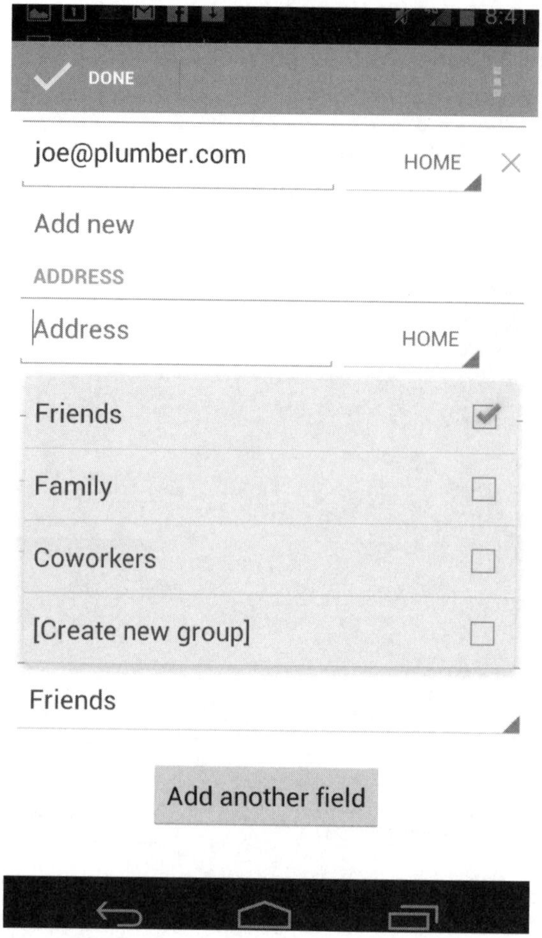

FIGURE 11.29: The Contacts app uses a smaller sole Add Another Field button in the middle of the screen.

Provided the sole action button is sized to at least 30 to 50 percent of the width of the screen, the central placement is both attractive and ergonomically satisfying, which makes the design in Figure 11.29 a good choice.

Pet Shop Application

The Pet Shop app follows the Trulia model to make the primary action button Register Pet stand out with both size, color, and a check mark icon. Refer to Figure 11.4 to see how this looks in a hand-drawn wireframe.

Tablet Apps

This pattern applies to tablets with the caveat that the entire button group be placed along the right edge of the device for easy accessibility without compromising hand position on the tablet. The example from the Calendar app in Figure 11.30 includes a good version of the recommended implementation, even though the buttons are reversed (OK/Cancel instead of Cancel/OK).

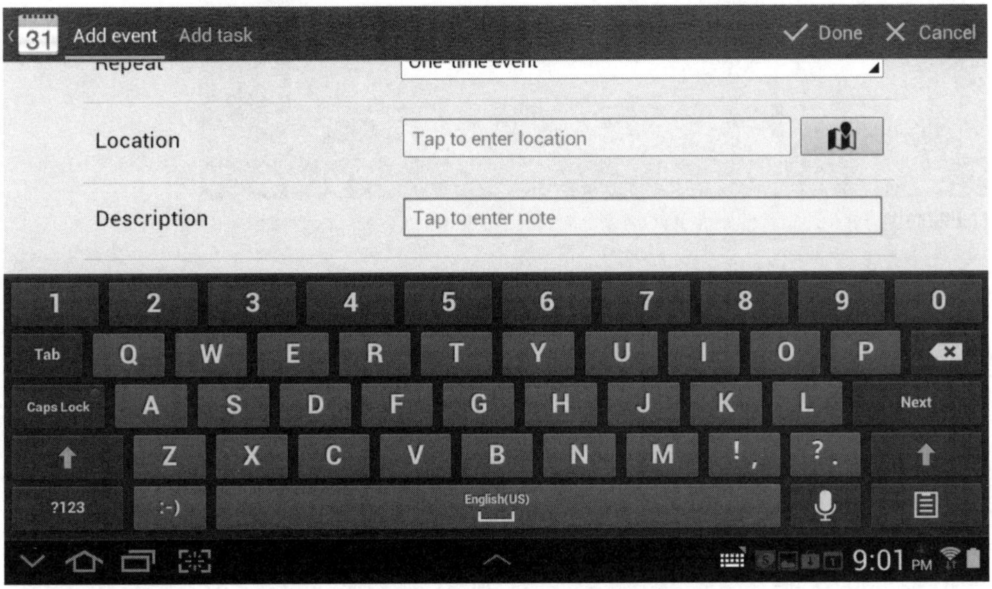

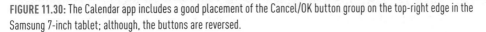

FIGURE 11.30: The Calendar app includes a good placement of the Cancel/OK button group on the top-right edge in the Samsung 7-inch tablet; although, the buttons are reversed.

Provided the action buttons are near the edge of the screen, the top placement is preferable to placing action buttons at the end of the form because of the better ergonomics. Top placement buttons are easier to find and enable optional form fields to be ignored without having to scroll all the way to the bottom of the form. The information in the "Caution" section does not apply to tablets. Unlike the smaller mobile devices, tablets are held in two hands, so there is no hand contortion needed to tap the top action buttons.

Unfortunately, tablet apps often ignore both ergonomics and conventions and place action buttons in a haphazard manner, which is an antipattern, as the lightbox in Figure 11.31 demonstrates.

FIGURE 11.31: The haphazard placement of the action buttons in the Calendar app on a 7-inch Samsung Galaxy Tab is an antipattern.

Even though the lightbox is fine, the action buttons are in the wrong order. The OK button is far away from the fingers of the right hand (which is the dominant hand for the majority of the population), and both buttons are presented in the same size and visual treatment. Set is a smaller word label than Cancel, which means that the primary action button is further de-emphasized! (The label "Set" is also unexpected; the more standard label for this action would be "OK.") Finally, note that the Cancel button is also much too close to the scrolling calendar, suggesting that it might be hit by accidental swiping. Altogether, this "antipattern smorgasbord" creates a strong possibility that people will often tap the Cancel button by mistake instead of tapping the Set button and thus will be frustrated several times a day—each time they need to add an appointment.

! Caution

Should you place actions at the top or at the bottom? *It depends.* The action buttons positioned at the top are easy to find, which is essential for a form like the New Event form in the Calendar app (refer to Figure 11.24) because most fields are optional, and in 90 percent of the cases, the customer will not scroll to the bottom of the screen.

However, obvious placement comes at a price of ergonomic accessibility: Buttons at the top are difficult to tap with one hand. This is both good and bad. Top placement is *bad* because most people expect to operate their device one-handed. For precisely the same reason, the top placement of action buttons is also *good* because the action buttons are out of the way and are unlikely to be tapped by accident.

For a one-time registration form, top placement is a good choice because you want your customers to be careful while filling out the form. However, for the Calendar app's New Event form, it's a bad choice because people want to comfortably use one hand to accomplish routine tasks like entering an appointment. Then why are the buttons at the top? For the Calendar, the optional form field requirement wins out, so the action buttons get top placement; placing them at the bottom of the form would require a lot of scrolling every time your customers fill out the form. For the Calendar app's New Event form, having the customer contort his hand to tap the top action button is the lesser of two evils.

Should you disable the primary action button until the form is filled out correctly? *No.* Disabling the primary action button is usually not recommended unless the form is simple or uses a Callback Validation pattern. If the action button is disabled in a long form and something that's not obvious is missing, the customer will need to hunt around the screen. That puts him at risk of not finding whatever it is that is missing/incorrect, which increases the chance that he will abandon the form altogether from sheer frustration. For example, if the action button were disabled in the Yelp form shown in Figure 11.32, the customer may never find out that he is missing the required Picture field, as it is nowhere labeled as required. And furthermore, pictures are not usually required in forms.

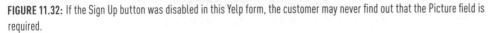

FIGURE 11.32: If the Sign Up button was disabled in this Yelp form, the customer may never find out that the Picture field is required.

However, because the action button is available and active, the customer may try to submit the form and get the error, which facilitates the corrective action. The result is a more robust and customer-friendly experience. In that case, even this rudimentary Pop-up Alert is better than no error reporting, which is what the disabled action button is tantamount to. (Recall that you should not use the Pop-up Alert pattern Yelp uses for reporting form errors—use Inline Error Message instead (see the "Caution" section for "11.3 Pattern: Pop-up Alert"). However, if you decide to ignore this warning and go with the disabled primary action button in your form, at least make sure you do some customer testing of the finished interface.

Should you make the primary action stand out compared to the secondary action? *Absolutely.* The recent trend has been to make both Cancel and OK look the same, as in the Kayak app refine screen shown in Figure 11.33. This is an antipattern.

FIGURE 11.33: Antipattern: Cancel and OK look the same in the Kayak app refine screen.

This is a bad decision because customers rely on subtle visual cues such as color saturation, size, text, and placement to make the choice of which button to tap. Do help your customer out, if you can, by making it *absolutely obvious* which button to push. As Steve Krug so famously quipped, "Don't make me think!" in his seminal book by the same title (2005, New Riders). This advice goes double for small mobile screens, where text on the button is often difficult to read in bright sunlight or because the device screen vibrates while the user walks or rides in a vehicle.

Related Patterns

11.1 Pattern: Inline Error Message

11.6 Pattern: Top-Aligned Labels

When it comes to labels, top-aligned is the standard implementation.

How It Works

When the form is presented to the customer, form labels appear above the fields.

Example

The Android reference implementation is the Calendar app's Add Event form, which is a frequent example in this chapter. (See Figure 11.34.)

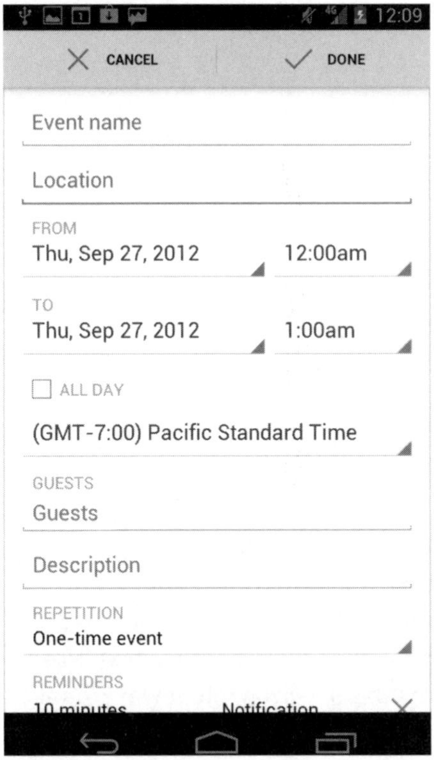

FIGURE 11.34: The Calendar app implements the Top-Aligned Labels pattern.

Calendar uses the standard Android 4.0 Ice Cream Sandwich OS visual scheme, so the labels appear in a smaller uppercase font, whereas the in-field Input Mask appears in a slightly larger mixed font. This can create a bit of confusion because everything runs together, especially on mobile devices with smaller screens.

A more OS-agnostic presentation of labels is on the eBay app registration form, as shown in Figure 11.35.

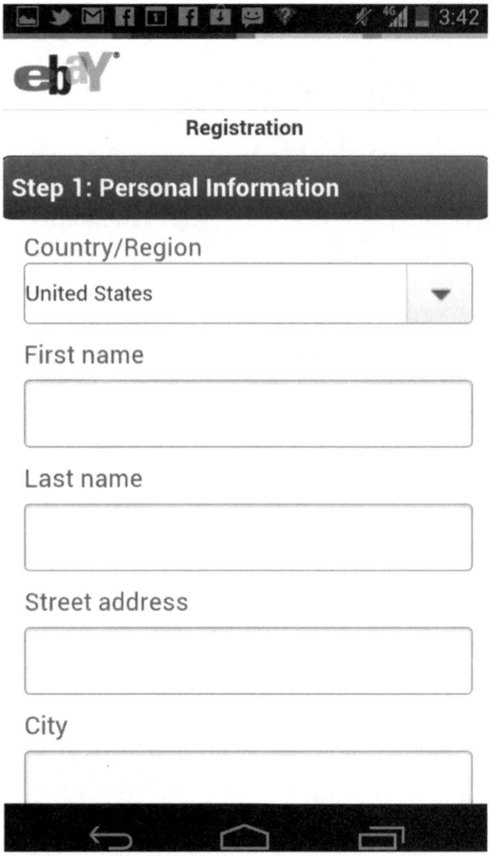

FIGURE 11.35: The eBay app uses a device-agnostic style for the Top-Aligned Labels pattern.

eBay's form fields are actual full boxes, which separate from the field labels better and create an overall cleaner user interface than the reference Android OS implementation. However, that is solely my opinion. What is a well-researched fact is that mixed case text offers better readability than all uppercase text. Presumably this insight applies equally well to form labels, so you might want to take that into account.

When and Where to Use It

Any time you present your customers with a mobile form, you should use top-aligned labels.

Why Use It

According to Luke Wrobleski's research that he presented at Design 4 Mobile, Chicago 2010 (http://www.lukew.com/presos/preso.asp?23), which is supported by my user research, top-aligned labels offer the best all-around versatility and usability on mobile.

Here's how top-aligned labels compare with other types of labels:

- **Left-aligned labels:** Compared with the top-aligned labels, left-aligned labels limit the size of the readable portion of the field, as shown in the form from Southwest in Figure 11.36. Note that even a modest-length e-mail address does not fit completely into the input field.

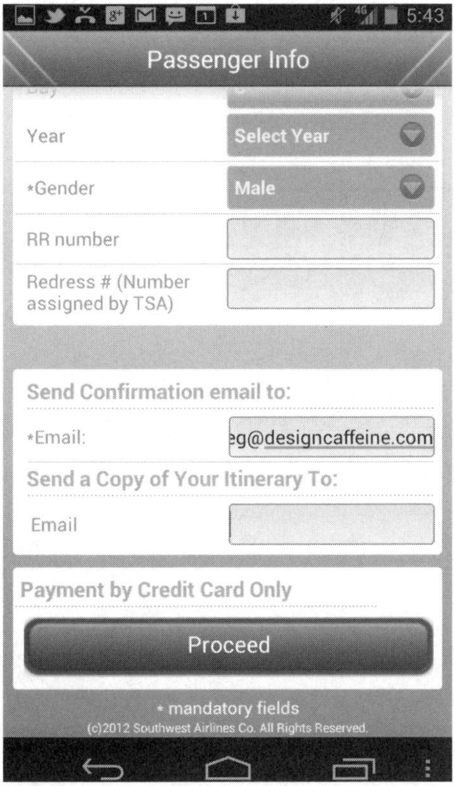

FIGURE 11.36: The left-aligned labels in the Southwest app limit the visible portion of the field.

- **In-field labels:** In-field labels are another popular alternative to the Top-Aligned Labels pattern. In-field labels generally work fine for short forms, such as Login, but they make it easy to get lost in a long form.

The problem is that the label is lost as soon as the person starts to edit the field, so the current field your customer is on and *all* the previously filled-out fields no longer have any labels. This is not the best option on mobile because if the customer is interrupted or distracted (as happens a great deal on mobile) she will not be likely to remember what the field was that she was in the process of filling out, and she will be confused—or worse, she'll abandon the form altogether. When the customer is editing the textbox with the in-field label, it becomes a plain textbox, as shown in the sign-up form from PayPal in Figure 11.37. Can you tell which field the customer is filling out in the form?

FIGURE 11.37: The in-field labels in the PayPal app make it hard to tell what field is being filled out.

In-field labels also do not allow the input format mask to be used, so it sharply limits their usefulness for more complex data entry that requires formatting.

In contrast to other labeling strategies, top-aligned labels do not have any of these inherent limitations. They use the entire width of the mobile screen; have complete flexibility to employ any format mask; and remain labeled throughout the form's life cycle. This makes top-aligned labels the best all-around choice for mobile forms.

Other Uses

You can also use top-aligned labels to identify the entire group of related controls, such as in the implementation from the Contacts app shown in Figure 11.38. In this case, the blue Events group label (plus a blue underline) delineate a group of two events.

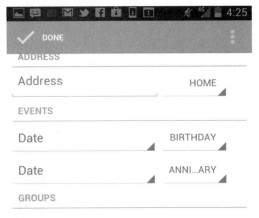

FIGURE 11.38: In this implementation, the Top-Aligned Labels pattern is used to label a group of Events fields in the Contacts app.

Pet Shop Application

Throughout this book, the Pet Shop app designs mostly follow the standard Android 4.0 Ice Cream Sandwich OS visual design guidelines for fields and their labels. Refer to Figure 11.4 to see how this looks in a hand-drawn wireframe.

Tablet Apps

Tablet apps enjoy ample real estate, so the Top-Aligned Labels pattern is not a requirement for tablets. However, it is easier to simply follow the mobile design conventions for forms, and that's exactly what most tablet apps do. Unfortunately, the vertical space is limited when the tablet is in the horizontal orientation, especially with the soft keyboard onscreen (see Figure 11.39).

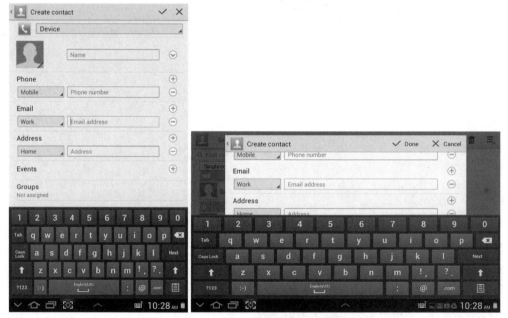

FIGURE 11.39: Top-aligned labels limit vertical space for tablets in the horizontal orientation.

One experimental pattern to try out for this situation is to make the labels fluid. When in the horizontal orientation, make the labels left aligned; when in the vertical orientation, make the labels top aligned. Figure 11.40 is a hi-def wireframe of what such a form might look like (compare it with the Figure 11.39). Because I haven't seen this customization, it remains an experimental pattern.

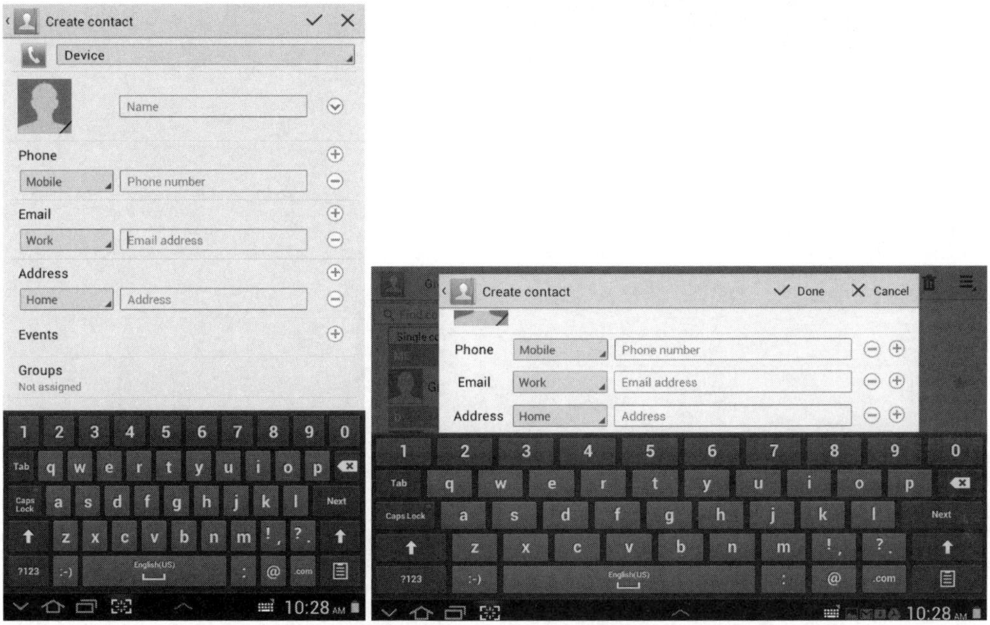

FIGURE 11.40: In this experimental pattern, labels adjust to optimize space based on the tablet's orientation.

! Caution

When using the Top-Aligned Labels pattern, you must be aware of the vertical space taken up by the label. This is particularly important for the group labels mentioned in the "Other Uses" section because those labels take up an additional 36 pixels of vertical space (see Figure 11.41).

FIGURE 11.41: A group label takes up 36 pixels of vertical space.

Thirty-six pixels does not seem like a lot, but it adds up quickly for long forms. Also, group labels add a vertical line, which could confuse things a bit with the new Android 4.0 visual scheme, which is almost entirely composed of horizontal lines. Avoid using group labels for simple forms. (See Figure 11.42 for an example of an antipattern of group labels in the Contacts app.)

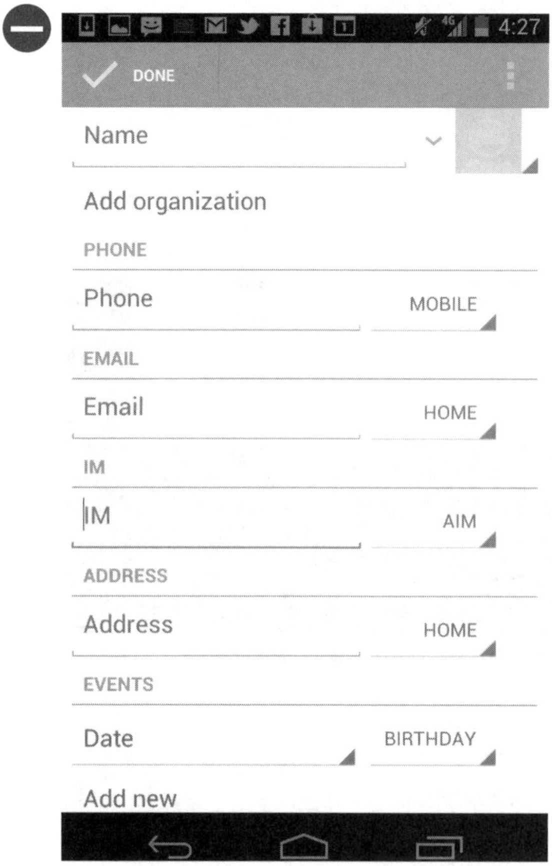

FIGURE 11.42: Using group labels for nongroup fields is an antipattern in the Contacts app.

The entire form in Figure 11.42 would be much cleaner and save approximately 180 pixels (5 group headings × 36 pixels each = 180 pixels) of vertical space if it used the standard single-field labels. You can always create groups dynamically, as the customer fills out the form, and update from single-field labels to group labels as needed—for example, if the customer adds another phone number, thereby turning a single field into a group of similar fields.

Related Patterns

None

11.7 Pattern: Getting Input from the Environment

Desktop forms are keyboard-centric and get little information from the environment. In addition, environmental information is not that useful; most desktop web forms are filled out in the office or at home, so the environment remains static and yields little useful information. In contrast, mobile forms are frequently filled out on the go, so they can often benefit from surprising amounts of environmental data collected with a mobile device.

How It Works

Whenever the form needs to be filled out, the mobile device uses the read out of the on-board sensors (voice, gestures, accelerometer, location, images, video, and ambient light) as form input.

Example

Chapter 7, "Search," describes at length how voice can be used as an input mechanism. Gestures as input are used often in games such as Angry Birds, Fragger, Hungry Shark, Grabatron, and Burn the City—although you rarely consider game inputs to be forms. (Although of course that's exactly what they are.) Figure 11.43 shows a welcome animation that explains the gestures for putting everyone's favorite angry bird into explosive action.

FIGURE 11.43: Gestures are used as input in the Angry Birds game.

Gestural input doesn't need to be limited to games, however. Map-based inputs are fairly common, with the on-screen map area (located by pinching and stretching gestures) used to constrain the corresponding list in the example from Trulia shown in Figure 11.44. Zooming out on the map presents a longer list of homes, going from three (in the screens on the left) to seven (in the screens on the right).

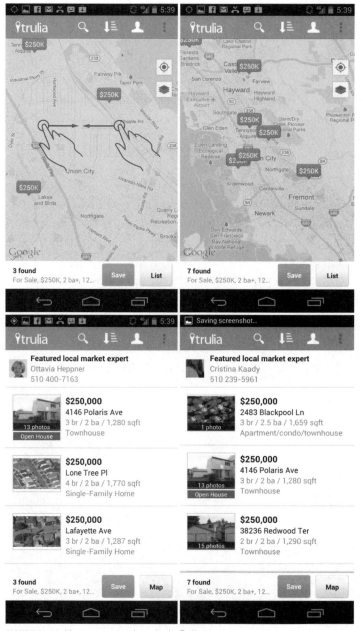

FIGURE 11.44: Map gestures are input in the Trulia app.

Gesture-based character input has a long history on touch devices, starting with Graffiti apps in the early versions of the Palm Pilot. The latest iteration of the original Graffiti input method is the Gesture Search app from Google Labs. Gesture Search enables simple phone operations and shortcuts to be performed by drawing letters and symbols using the entire screen as a touch pad and the finger as a stylus. The interface also enables customers to erase letters by drawing a horizontal line. (See Figure 11.45.)

FIGURE 11.45: Gesture Search uses sophisticated gestural form input.

Another popular option to get input from the environment is to use Location as an input. As Luke Wroblewski reported in his Design 4 Mobile 2010 workshop, location can be obtained by the mobile device through many means, each of which has different performance characteristics, such as speed, precision, and effect on the device battery. Although the mobile industry usually abbreviates the entire bundle of location services as "GPS", Global Positioning Service (GPS), the satellite navigation system, is only one of the ways the phone obtains location information. It is the most accurate method; it can locate the mobile device to approximately 33 feet, but it can take anywhere from 2 to 10 minutes to obtain the position. Also, using GPS drains the battery quickly.

Another popular location service is triangulation using multiple cell towers. Cell tower triangulation has a negligible effect on the battery and provides nearly instant location information to a range within 1,600 to 8,200 feet for a single tower and 328 to 4,600 feet for multiple towers.

Most mobile devices obtain location information via a combination of both methods (plus Wi-Fi, when available). When location function is first called, most people observe the phone drawing a large circle initially based on the cell tower triangulation. If GPS is enabled, the location circle gets progressively smaller, and the location gets more precise in a few minutes as the GPS comes online. Location is a favorite input for mobile devices because it is useful for local in-person interactions, such as those facilitated by Yelp (see Figure 11.46).

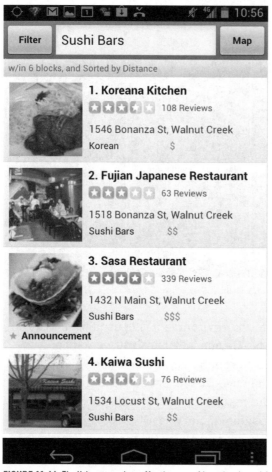

FIGURE 11.46: The Yelp app makes effective use of location-based form input.

Image-based input is becoming increasingly common as well, particularly to take a picture of a QR code. QR codes can have a variety of payloads, depending on the size of the code and encoding density—from a chunk of text nearly the size of the American Constitution (4,296 characters to be exact) to phone numbers and URLs, all of which can be used as simple form inputs. QR codes can also encode a variety of specialized formats, such as MeCARD, which contain complete structured data that can be interpreted for the purposes of filling out a form. Out of the many QR code readers currently on the market, Red Laser remains one of the easiest to use and the most versatile (see Figure 11.47).

FIGURE 11.47: The Red Laser QR code reader is in action reading a business card with a URL payload.

READ MORE ABOUT QR CODE UX STRATEGIES

You can read more about the use of QR codes on this book's companion site where I host a series of six QR Code UX strategy articles and a virtual seminar on the topic "QR Codes That Convert: Mobile UX Strategies for Success," April 6, 2012. To navigate to this bonus content, simply scan the QR code in Figure 11.47 or point your browser to http://www.androiddesignbook.com/qrcode/.

But do you need encoding to use images as form input? Not necessarily. One brilliant example of using simple pictures to establish the connection between physical and virtual is the Amazon Remembers feature of the Amazon.com mobile app. (See Figure 11.48.) Amazon Remembers uses a combination of the optical recognition and Mechanical Turk service to interpret pictures of most items that can be bought on Amazon.com. All the customer needs to do is take a picture. The picture is then seamlessly uploaded to Amazon.com, often coming back within seconds with the identity of the item and access to price, options, description, and, of course, reviews and the ability to buy the item or add it to your wishlist, truly connecting the physical item and its virtual informational footprint.

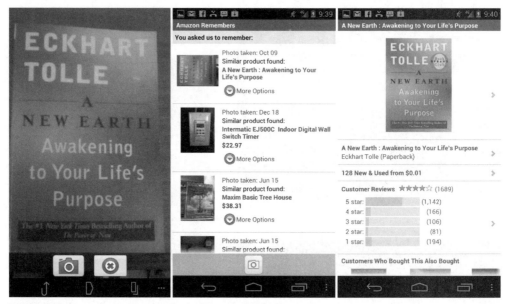

FIGURE 11.48: Amazon Remembers uses simple pictures as form inputs.

Brilliant services like Amazon Remembers give you a hint of things to come in the field of the Internet of Things that Bruce Sterling talked about in his amazing book *Shaping Things* (2005, The MIT Press). The term Internet of Things was first used by Kevin Ashton in 1999 and refers to uniquely identifiable objects (things) and their virtual representations in an Internet-like structure. The Internet of Things is a dynamic global network infrastructure with self-configuring capabilities based on standard and interoperable communication protocols, where physical and virtual "things" have identities, physical attributes, and virtual personalities; they use intelligent interfaces, and they are seamlessly integrated into the information network (http://en.wikipedia.org/wiki/Internet_of_Things).

When and Where to Use It

Any time you have a form, you should be immediately thinking, "What inputs can I get from the environment?" Can you use gestures, location, pictures instead of the keyboard, and native input controls?

Why Use It

Environmental inputs are not just more fun than regular inputs. They are functional as well, providing more efficient and effective input strategies uniquely suited to mobile devices.

Other Uses

Getting input from the environment does not need to be done one alternative technology at a time. Using a combination of various methods can also be effective. For example, Yelp gives an excellent demonstration of the combination of accelerometer, GPS, and video used as inputs in its augmented reality feature, Monocle (see Figure 11.49).

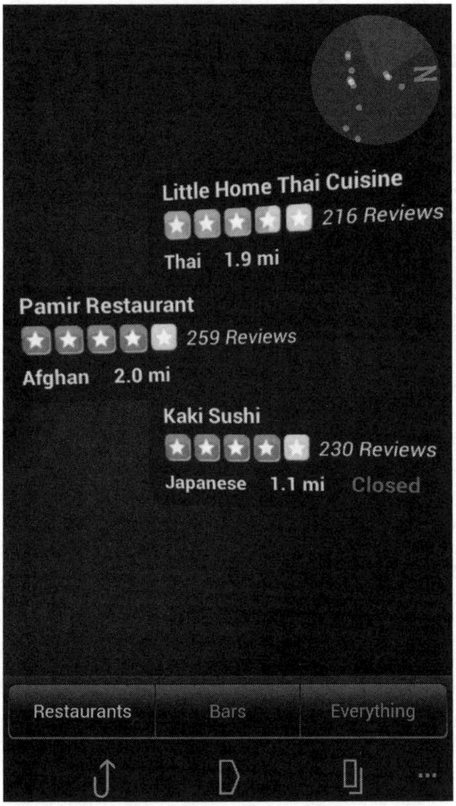

FIGURE 11.49: Yelp's Monocle demonstrates the effective use of location, accelerometer, and video form inputs.

Likewise, don't be shy about requesting form input help from external resources, such as remote virtual assistants. Apps such as YCard use a Mechanical Turk remote assistant service to interpret image-based input of business cards. The customer takes the image of a business card and then remote virtual assistants read it using a combination of Optical Character Recognition (OCR) and human eyes to input the information into a standard desktop form. The entire process is quite fast; it takes only 30 minutes or less. As you can see from Figure 11.50, the app even reads Chinese characters and translates them into English. This is a classic use of a Virtual Assistant pattern mentioned in Chapter 7 that uses images as input.

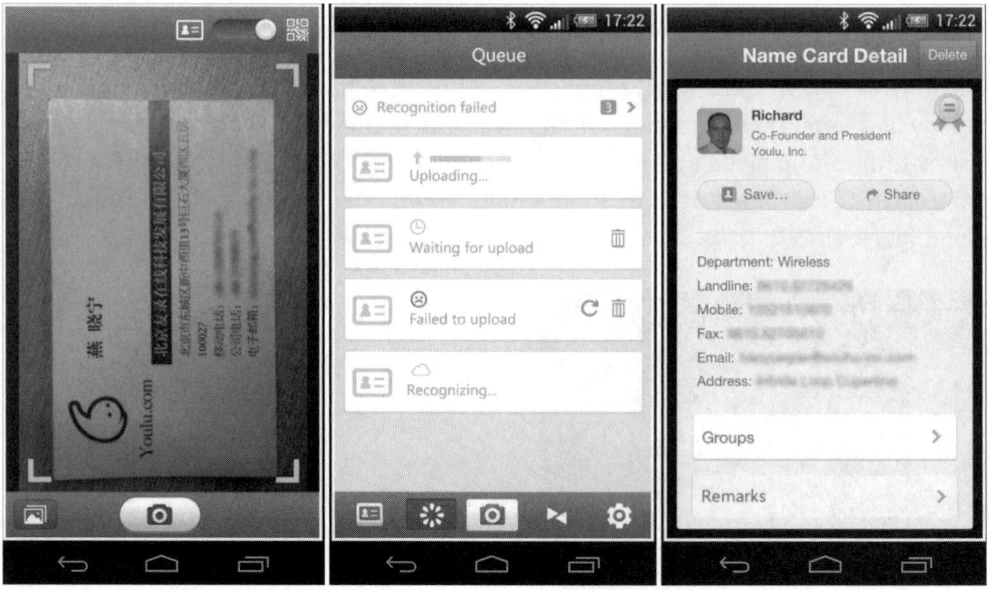

FIGURE 11.50: YCard uses image form input in combination with Mechanical Turk services.

Other OCR/Mechanical Turk services are used in financial apps to do everything from depositing checks to compiling and categorizing expense receipts.

Pet Shop Application

In the Pet Shop app example, when the customer takes a picture of his pet, in addition to storing the picture, the app will attempt to use optical recognition technology to determine what kind of pet it is (see Figure 11.51). For example, a "simple" learning program can recognize whether this is a picture of a dog and a more sophisticated algorithm can figure out the breed and approximate age of the pet, making data entry easier.

FIGURE 11.51: The pet registration form uses image as input in combination with a Mechanical Turk service.

Developing a service like this is definitely not trivial; there are probably only a handful of people in the world who could code an algorithm with the precision necessary to reliably tell, for instance, a Labradoodle from its constituents, Labrador and Poodle. However, the Mechanical Turk model can work reasonably well, provided the app workflow affords a built-in delay that enables a remote virtual assistant to review the picture and perform identification and data-entry work. As Tim Ferris mentioned in his book *The 4-Hour Workweek* (2009, Harmony), companies such as Brickworks India offer relatively inexpensive remote virtual assistant services that can be configured on the back end as a Mechanical Turk for all kinds of applications.

A better question to ask would be whether it is worth making the pet owner wait 30 minutes to enter three pieces of data at further risk of offending the customer by guessing the breed incorrectly. (Or worse, saying something is a cat when it's actually a dog. Imagine the insult!) This rather contrived example points out some limitations of the pattern. However, for many mobile applications, outsourcing complex form data entry to a Mechanical Turk Virtual Assistant service holds a great deal of promise. Just make sure that the "juice is worth the squeeze."

Tablet Apps

You can use tablets in the same way as mobile devices. One of the biggest complaints that customers had of the Chase iPad app (one of the first financial apps made specifically for a tablet) was that it could not deposit checks remotely by taking a picture of them in the same way that the Chase mobile app could do. The firm customer expectation for tablet apps seems to be that they will do *more*, not less, than their mobile counterparts.

! Caution

The earlier Angry Birds example underscores an important point: When using gestures, provide clear, actionable instructions on how to perform them. The best way to do that is to use a Watermark pattern, which is covered in Chapter 13, "Navigation."

Related Patterns

7.1 Pattern: Voice Search

13.5 Pattern: Watermark

5.5 Pattern: Tutorial

11.8 Pattern: Input Accelerators

The previous section about getting input from the environment discusses getting data *into* the mobile device using a variety of on-board sensors. This section discusses ways to *retain* data between visits and present this data at strategic points of the workflow to re-engage the customer in the previous task.

How It Works

When the customer performs complex, time-consuming data-entry tasks such as entering dates and addresses, the system remembers what was entered and surfaces it at the proper time during a future engagement.

Example

The Maps app is an excellent all-around app that is made even better by using Input Accelerators. When the Maps app launches, it displays the map of the immediate area. Tapping the search function brings up a search box with the auto-suggestion layer prepopulated by the previous search requests (see Figure 11.52).

This is extremely handy because the addresses are hard to type and difficult to recall, yet a simple act of navigation to your destination is likely to be interrupted multiple times with tasks like calling, texting, and music listening. In addition, many of the destinations a customer visits are likely to be visited more than once, as are some common queries like "Coffee" or "Starbucks." Finally, the layer offers the Tap-Ahead pattern—the ability to enter the query into the search box but retain it in the editable mode.

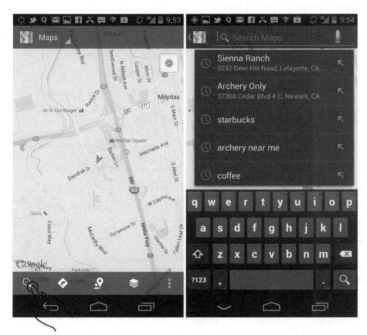

FIGURE 11.52: The Maps app includes an excellent auto-suggest layer Input Accelerator.

When and Where to Use It

Any time your customers might want to redo the same complex data entry more than once, you should use this pattern. Look particularly for dates, addresses, city names, contacts, titles, descriptions, and other similar data entry fields.

Why Use It

Entering anything on a mobile device is hard. Forcing your customers to not only remember but also *re-enter* the same data is downright evil. Don't be evil.

Other Uses

On a related note, the most important accelerator you can provide is to *ask people to enter less information.* The Yelp sign-up form discussed earlier in the "11.3 Pattern: Pop-up Alerts" section makes a great negative example. It never makes clear which fields are required, until the customer scrolls to the bottom and sees the Gender (optional) field. Also surprising is that the Yelp sign-up form requires a picture, which is highly nonstandard. No matter how many accelerators you put into your forms, nothing affects the form completion rate as much as having extraneous fields. Remove extra information requests, and wherever possible use smart defaults and direct manipulation controls.

Pet Shop Application

An open auto-suggest layer is not the only way to implement Input Accelerators. Sometimes, you need to give your customers an option to recall field values individually. A common example is that of planning a trip—a typical customer would want to try several different combinations to find the best overall value. For example, a customer may initially want to fly in on Thursday morning and fly out on Sunday night. However, Sunday night flights may be overbooked, so she might want to try returning on the preceding Saturday or next Monday instead, keeping the original Thursday departure date. Input Accelerators make this sort of exploring much easier. One way to implement them is shown in Figure 11.53 in the hand-drawn wireframe for the Pet Shop Travel app (Best Vacations for You and Your Pet).

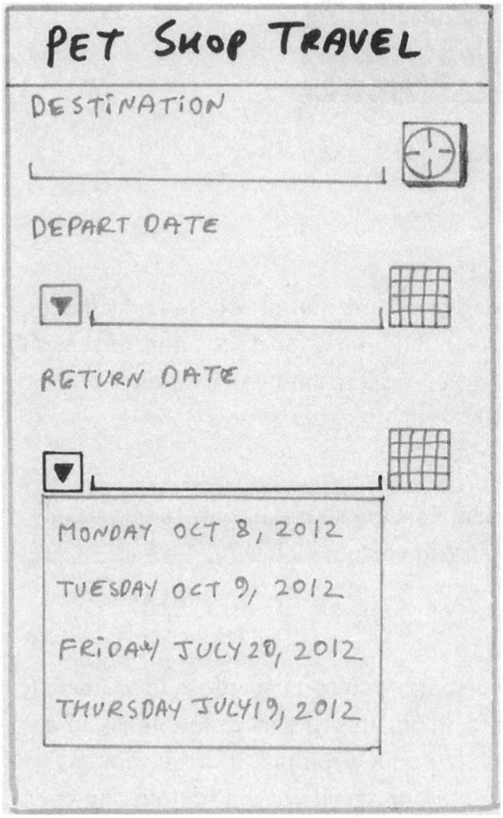

FIGURE 11.53: This experimental form of the Input Accelerator pattern uses a drop-down in the Pet Shop app's travel date inputs.

The wireframe in Figure 11.53 implements individual Input Accelerators on both the Departure and Return date fields via a drop-down that's hidden under a down arrow in front of the field. If the customer wants to recall one of 10 previous values, she would just tap the arrow to get the drop-down layer showing the previously entered values. Putting the arrow in front of the field reinforces the subtle suggestion of looking there first and keeps the input accelerator safely away from the calendar button. In this way a great number of useful date combinations can be explored quickly, and a date doesn't need to be entered more than once.

Tablet Apps

Tablets have plenty of real estate to show previous values. There is literally no excuse not to show previous search results as an Input Accelerator on a tablet.

One great option is to share the cache data across multiple devices, as the Maps app does on the Android 4.0 platform (see Figure 11.54).

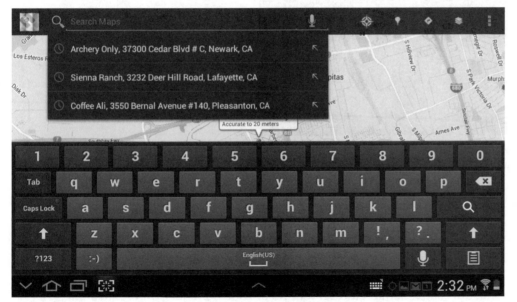

FIGURE 11.54: The Maps app shares Input Accelerator values across devices.

Do keep in mind the device orientation. Even though the Maps app screen in the vertical orientation on the tablet has a lot of space, the width of the auto-suggest is too small to read the values comfortably (see Figure 11.55).

FIGURE 11.55: The width of the auto-suggest Input Accelerator is too small to accommodate most values in the vertical orientation.

! Caution

With any great power comes great responsibility. Avoid caching sensitive information such as credit card numbers, unless your app specifically advertises this feature. When recalling details like bank account and credit card numbers, it is best to cache them on the server and provide a 2-Factor Authentication (see "12.1 Pattern: Login Accelerator" in the next chapter) to access the data.

Remember to provide a way to clear the previous entries cache on the device for the entries that your customer wants to keep private. I haven't found a setting that clears the Map app cache, for example, or controls cross-device sharing.

Related Patterns

6.6 Pattern: History

CHAPTER 12

Mobile Banking

This chapter departs from the format of the book somewhat. So far the book has referred heavily to e-commerce, social media, lifestyle, productivity applications, and games. This chapter focuses specifically on the rapidly emerging field of mobile personal finance and mobile banking. In keeping with the Pet Shop theme, you can imagine you formed your own Pet Shop Bank—maybe to save for Fido's retirement? Although these patterns are placed in the context of mobile banking, these same patterns and considerations apply just as readily to any long forms and complex workflows that deal with high-stakes data entry in a security-conscious environment.

12.1 Pattern: Login Accelerator

The Login Accelerator pattern enables accelerated mobile login using a short code, facial recognition, or voice imprint while providing an acceptable level of security.

How It Works

Some banks offer the customer the option to add the device to the list of "approved" devices by installing a special code on the device in a form similar to that of a browser cookie. The customer also sets up a second level of authentication in the form of a four - to six-digit numeric code, a picture of his face, a voice imprint, or some other mechanism that involves less typing than the typical username/password combination.

Example

To set up an accelerated login, some mobile banking apps, such as Chase (see Figure 12.1), force the customer to set up *two-factor authentication*. The first part of the authentication (the first "factor") is the login and password to get into the Chase app; the second part of authentication is the access to the e-mail or a phone already registered to the customer, verified via a one-time setup token that is e-mailed or sent via SMS.

FIGURE 12.1: The Chase mobile banking app uses two-factor authentication.

After the code is received and accepted by the device, and the two-factor authentication is complete, the mobile phone device becomes "authorized," and the person can use the app for mobile banking. To accelerate the login, the app can also remember the username to save the customer the time and effort of typing it each time he wants to access the app. The user tells the app to remember the username by tapping the Save User ID check box on the login screen, as shown in Figure 12.2.

FIGURE 12.2: The Login Accelerator in the Chase app saves the User ID.

In contrast to Chase, the USAA bank app enables the customer to set up a Login Accelerator using a combination of username, password, and his existing PIN without waiting for external confirmation by e-mail or text. After the initial authentication, the customer has the option to enable login using just his four-digit PIN simply by going to the Setting screen—note the instructions to that effect that appear in a faux metal box on the login screen (see Figure 12.3).

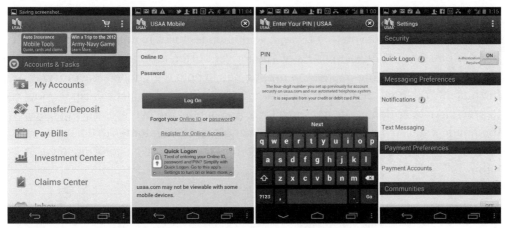

FIGURE 12.3: The USAA banking app uses a Login Accelerator with a four-digit pin called Quick Logon.

True to its name, after the Quick Logon is turned on, the customer can access the app quickly just by entering the four-digit pin (see Figure 12.4).

FIGURE 12.4: After the initial setup, the Login Accelerator in the USAA Banking app allows customers to login quickly.

The method to unlock the app is not limited to a four-digit code. The Android 4.0 Ice Cream Sandwich Nexus phone features a facial-recognition phone unlock mechanism, which is accurate and impressive.

After the customer sets up the mug shot of his face under various light conditions and angles, he can activate the facial recognition feature. To unlock the phone, all the customer has to do is…turn it on. The Power button is located in the top-left corner, so to turn on the phone, the person has to hold the phone a certain way, facing him. The built-in, front-facing camera is positioned at just the right angle to start taking a video of the person's face. As soon as the owner is recognized, the phone unlocks automatically, without requiring the customer to do anything further (see Figure 12.5). In the words of Peter Morville, the design "dissolves in behavior."

FIGURE 12.5: The Login Accelerator pattern in Android 4.0 Nexus uses facial recognition.

When and Where to Use It

Any time you have a login and username, consider whether you want your customer to type them in every time she needs to access her data. Whenever you can, give your customer the option to store the password and username on the phone for medium-security apps such as social networking, lifestyle, and personal information management. For apps such as e-commerce, delay the authentication for as long as possible, preferably until the checkout flow.

For frequent-use, high-security apps such as mobile personal finance, strongly consider using some kind of a Login Accelerator scheme to make login easier.

Why Use It

If logins are tedious on the desktop web, they are ten times more so on mobile, where tiny keyboards and fat fingers get in the way of typing a long e-mail username and password precisely. Often, the customer must switch the keyboard several times to type in special characters in the username and password—every time the app times out. In short, *logins make apps hard to use.* And to quote Steven Krug in *Don't Make Me Think* (2005, New Riders), "if people find something hard to use, they just don't use it as much." Login Accelerator enables a fast, secure mobile login that drastically lowers the login barrier.

Unfortunately, some of the worst offenders in this regard are pure HTML 5 apps such as Yahoo! Mail. Until recently, Yahoo! Mail was unable to save either the username or the password, forcing its customers to type 20 or 25 characters any time they needed to check e-mail, which could typically be 10 or 15 times a day. A recent update to the Yahoo! Mail app features a native implementation that automatically saves the username, but there is *still* no option available to save the password or accelerate the login in any way.

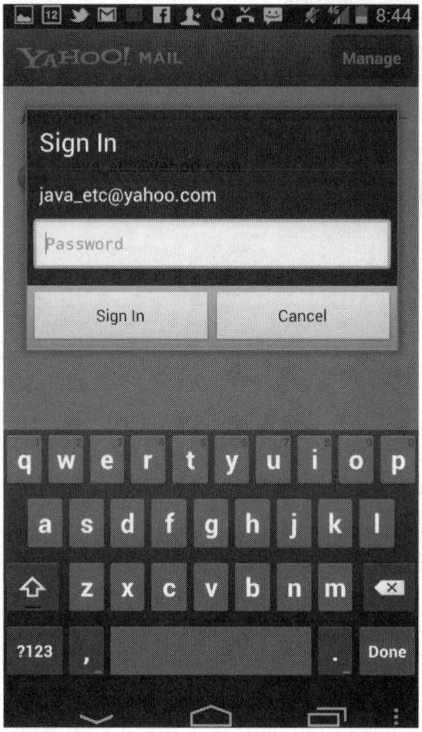

FIGURE 12.6: A lack of login accelerators in Yahoo! Mail makes it tedious and hard to use.

Speculatively, this inability to accelerate the login cuts down on the actual use of the Yahoo mail app. This is certainly true in my case.

Other Uses

One of the most intriguing ideas in recent years has been to use a physical token with near field communication (NFC) as a Login Accelerator. For example, imagine an NFC chip imbedded in an article of personal jewelry, such as a ring, used to unlock the phone on its own or in combination with some other form of authentication, such as a four-digit code or facial or voice recognition. This kind of a "decoder ring" would operate via an NFC chip and be more user-friendly than the RSA token many people use today to log in to a virtual private network (VPN). The security would be provided by the passive presence of the ring in the vicinity of the mobile device: If the ring is nearby, the mobile device detects the unique NFC signal and enables access. If the device is moved away from the ring (lost or stolen), the mobile device requires a more lengthy back-door access through a username/password authentication mechanism and alerts the owner to its location and a possible break-in attempt. NFC chips are now small enough to allow them to be placed in all sorts of small personal articles and even implanted subcutaneously. For the ultra-rich or people requiring this sort of permanent higher-level security, getting implanted with an NFC chip may be a worthwhile investment.

Pet Shop Application

Recall that you've already seen a four-digit code used instead of the password during the registration process described in the previous chapter in Figure 11.4. This current example pushes the envelope still further, with the voice recognition unlock mechanism. Imagine that during the registration the app asked the customers to set up the voice password in response to a simple challenge question, such as, "What is your favorite restaurant?" and the unique voice modulations of the owner were recorded to be used later as the means of identification. The app also sets up a four-digit code to be used as a backup.

When the app is launched, it immediately enters the voice recognition unlock mode, indicated by the oscilloscope display. The person speaks the password, and the app unlocks—no need to push anything more. In loud or quiet environments, the person also has the option to tap the Cancel button, and enter the four-digit backup code (see Figure 12.7).

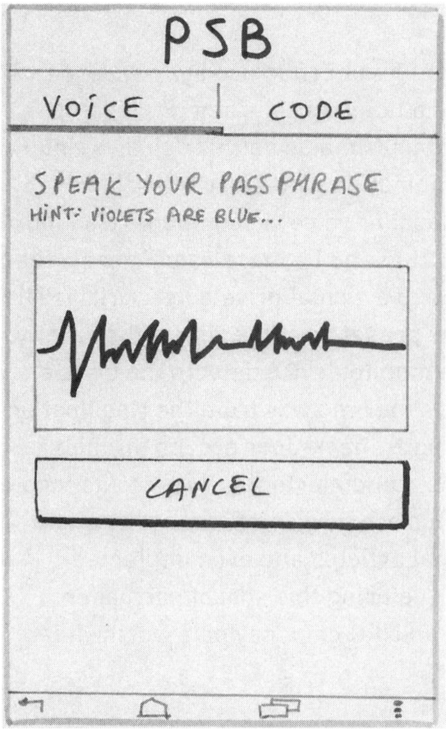

FIGURE 12.7: This experimental form of the Login Accelerator pattern could be used as a voice recognition password in the Pet Shop Bank app.

But why stop using voice at the level of the identification of the security for an individual app? As discussed in Chapter 7, "Search," voice-driven commands can be used with the help of a digital personal assistant (such as whatever Siri counterpart Google creates) to do routine tasks such as open apps. What if the app's name was used as a password and all you had to do was tell your phone to launch the app—the security of voice recognition handled directly by the Android OS? You can even get clever with the app's nickname, recognizing preset phrases such as "Show me Mack's accounts" (Mack being the name of the dog) to authenticate the voice modulations and the passphrase together to open the Pet Shop Bank app.

Obviously, unless Mack's retirement at a private tropical island was at stake, you don't need this level of security. It's included only as a demonstration and to make the point: High security does not need to be a hassle. This technology is already here and this "future" design pattern is already in use; it's just not, in the words of William Gibson "evenly distributed" ("The Science in Science Fiction" on *Talk of the Nation*, NPR, November 30, 1999).

Tablet Apps

Two-factor authentication provides a unique opportunity for tablet devices. As described in Chapter 17 of *Designing Search* (2011, Wiley), based on brilliant research done by Marijke Rijsberman, tablets, especially larger ones, are still expensive and rare enough to be considered "family devices." Although this may change in the future with every person on the planet having a tablet, the reality on the ground is that a large tablet is a *shared alpha device*, which means multiple family members take turns using it, much like the television sets in the 1950s or family PCs in the '80s and '90s. Unfortunately, the Android OS is not well suited to multi-person use because as of the date of this writing, it does not provide multiple logins as part of the operating system. Thus it is up to the designers of the individual apps to introduce a login or profile system.

It is both a challenge and an opportunity to create profiles in a family-friendly environment. One great model for this is Wii's Miis—a low-security, family-friendly profile system developed by Nintendo. Miis are styled to look like individual family members and retain gaming statistics and other profile attributes associated with individuals. Since *Designing Search* was published, Microsoft Kinect went a step further in identifying the individual family members by height and movement style, creating an amazing automatic Login Accelerator system that dissolves in behavior.

Because families often use large tablets for gaming, it makes sense to look at gaming systems for best practices in profile identification and management. Take the time to explore whether your app will be used by various family members. If so, consider providing facial or voice recognition, which could be as simple as speaking your name to activate your own profile and unlock the app. This is a simple system even a four-year-old can use.

Providing individual profiles makes particular sense for a shared large tablet because those devices are used primarily inside the home (where it's easy and comfortable to speak out loud your own name) and are rarely taken out of the building.

! Caution

Facial recognition is good when you are out in the public during the day in your civilian clothes. But what if you need to use your phone at night, in your bat-cave, when you are wearing your superhero disguise? (See Figure 12.8.)

FIGURE 12.8: Don't forget to provide back-door access.

Your customers may not all have lives as interesting as Bruce Wayne, but there are always circumstances in which the shortcut method of access set up through Login Accelerator does not work. Be sure to provide a back-door access that allows for the usual username/password combination in case they need to unlock the app "the long way."

Related Patterns

7.1 Pattern: Voice Search

12.2 Pattern: Dedicated Selection Page

When the customer needs to make a selection from a list, the Dedicated Selection Page (also known as a Full Screen Picker) is a great alternative to the Drop Down pattern, which you can read about in Chapter 10, "Data Entry."

How It Works

When the customer needs to select a value from a long list, she taps the selector control, and a dedicated page of values opens. After the customer selects the value on the dedicated selection page, the page closes and the system again displays the form with the selection shown in the selector control. This selection is often accompanied by the "sliding" motion transition left-to-right to the dedicated selection page and then right-to-left to get back to the form. This sliding motion transition led some designers to call this pattern "a slider" (not to be confused with the Android Slider pattern covered at the beginning of Chapter 10).

Example

One example of such a dedicated selection page comes from the Chase mobile banking app. To select a recipient for the bill pay app, for example, the customer navigates to the separate dedicated page where recipients are listed. (See Figure 12.9.)

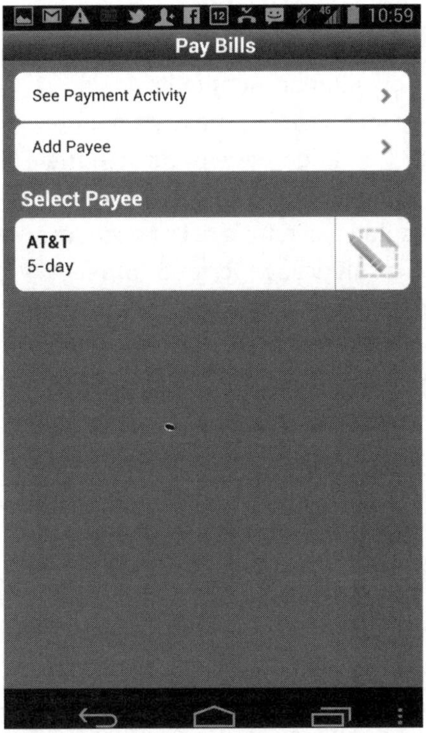

FIGURE 12.9: The Chase mobile banking app includes a Dedicated Selection Page for bill pay recipients.

After the customer selects the recipient, the system navigates back to the form page.

When and Where to Use It

Any time you have a large list to select from, consider using the Dedicated Selection Page pattern. This pattern is especially handy when you need to manage the list or have other controls such as a Scrubber to navigate quickly to the appropriate selection. Other considerations for using the dedicated page include customizing the dedicated selection page completely with colors, fonts, and styles of your choice. The simple drop down select control in Android enables only limited customization of the look and feel.

Why Use It

Selecting from a long list is a common task, as discussed at length in Chapter 10. As digital lives become more and more complex, lists of selections likewise get longer and more complex. The Dedicated Selection Page pattern enables a straightforward selection mechanism and empowers the designers to take advantage of the full screen width and full array of selection tools and navigation tabs.

A simple transfer of funds from a person's own checking account to her own savings account (known in the industry as a *me-to-me transfer*) makes a simple example to demonstrate the use of the Dedicated Selection Page pattern in the mobile banking environment. In the course of the me-to-me transfer, the Dedicated Selection Page pattern makes it easy to display a dynamic set of accounts and their values reliably without resorting to some funky JavaScript or other client-side coding languages. For example, if Checking Account A is chosen as a From account, you do not want to show Checking Account A as a possible selection for the To account. The Dedicated Selection Page pattern makes it easy to dynamically remove a value from the list or even perform more involved list manipulation according to complex business rules by taking a trip to the server before displaying the dedicated selection page.

Other Uses

It's easy to imagine this pattern extended to the Text Box with Atomic Entities pattern (refer to Chapter 10). For example, in one alternative design , the page that shows the entire list is also searchable via a simple search box on the top of the Dedicated Selection Page. This arrangement enables the Dedicated Selection Page to be browsed (by scrolling the list) as well as searched, similar to the way contacts can be browsed or searched in the Contacts app.

The Chase app (refer to the dedicated selection page in Figure 12.9) enables list management and displays the list. There is an additional button on the top of the list that enables a customer to add a new Payee if she does not already see her Payee in the list.

Pet Shop Application

Figure 12.10 shows how the pattern looks when it's used for a me-to-me account selection in the Pet Shop Bank app. The design implementation in the figure enables the customer to select her desired account by searching or simply by scrolling through the list of accounts.

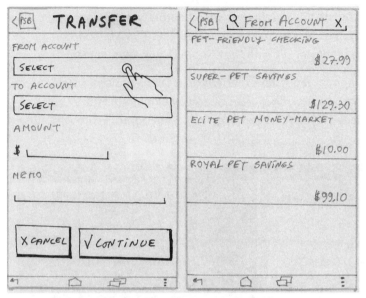

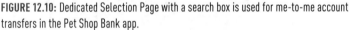

FIGURE 12.10: Dedicated Selection Page with a search box is used for me-to-me account transfers in the Pet Shop Bank app.

It also enables customers to easily limit the size of the list by typing in a few characters of the account name in the search box on the top of the page. The search box is fixed, and the account list scrolls below it. The keyboard does not display unless it is called for by tapping inside the search box.

Tablet Apps

For most tablet apps, the Dedicated Selection Page pattern is overkill. However, you can still use the cool features of the pattern in a lightbox pop-up or similar device. Unfortunately, most tablet apps do not take the time to do this; as a result the form interface quickly starts to look overly large and cartoony, especially on larger tablets. For more tips on implementing list selection on a tablet in both orientations, see Chapter 14, "Tablet Patterns."

! Caution

Don't overuse this pattern. If you have a simple static list of manageable size, without any dynamic additions, you do not need a Dedicated Selection Page pattern. A simple select can be an effective device that has a nice property of staying in the context of the current page (because it opens in a lightbox) without yanking the customer out of the flow completely, as a Dedicated Selection Page does. One thing to consider before using a Dedicated Selection Page is whether the strings in the list wrap two or more times when you use only the middle 80 percent of the page, as in a Select control. If the strings do wrap or you absolutely must have custom treatment of the selections, go ahead and use the Dedicated Selection Page pattern. Otherwise, you may be better off with the simple Drop Down control.

For example, the Chase app includes an excellent approach—it uses a Dedicated Selection Page for a more complex flow of selecting a recipient (refer to Figure 12.9) and a simple Drop Down select control for choosing the From Account (shown in Figure 12.11).

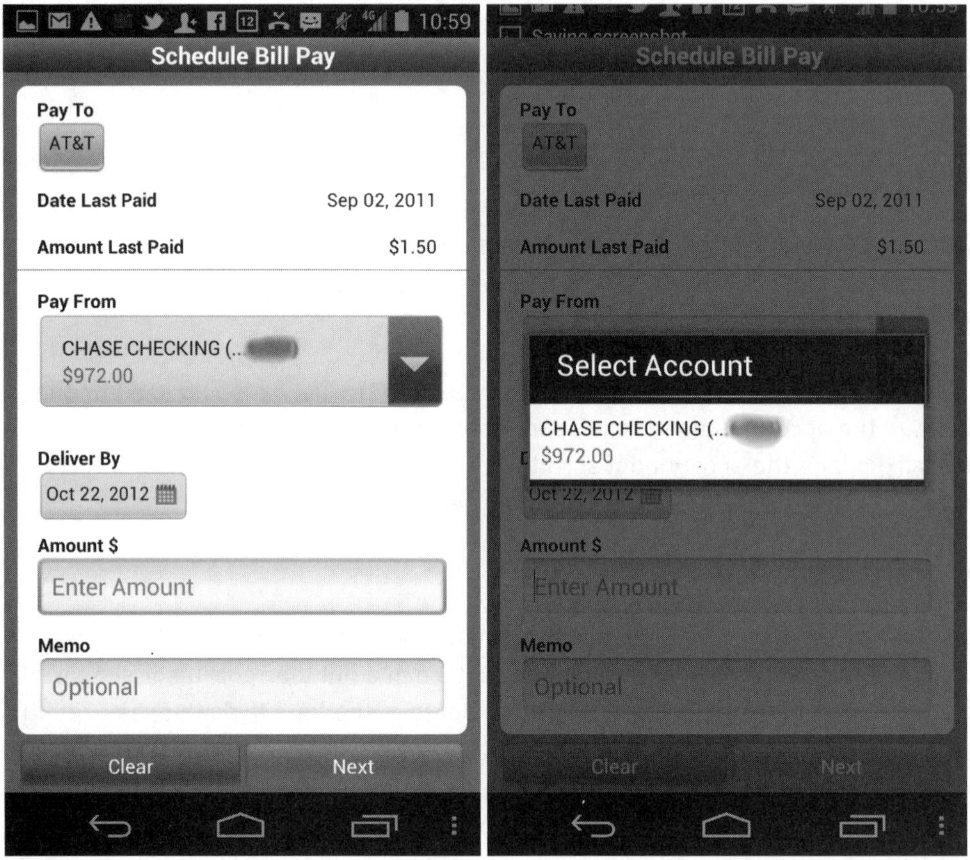

FIGURE 12.11: The Chase app uses a simple Select for the From account in the Bill Pay flow.

Related Patterns

10.5 Pattern: Drop Down

12.3 Pattern: Form First

When filling out a long form on the mobile device, many designers first try to copy the desktop web form. Although this has the advantage of being the most intuitive and obvious choice, this pattern typically results in longer flows that require more page views and more taps than alternative approaches in this chapter.

How It Works

When the customer needs to enter some data, he is first shown a form. All subsequent complex data entry screens return the person back to the form before going onto the next data entry.

Example

One example of this pattern is the USAA internal transfer—also called me-to-me transfer (see Figure 12.12). To enable transfer of funds from checking to savings, the customer is first shown a form with blank values. As the customer goes into the Dedicated Selection Pages (as discussed earlier in this chapter) the customer always returns to the form page but now with the appropriate account selections filled out.

With the verification page (discussed later in this chapter), the entire me-to-me transfer flow is *eight* screens as shown in the figure. Although this pattern generally works well for mobile devices, it's obviously a bit long; it's the longest flow pattern in this chapter.

When and Where to Use It

When your customers have a long form to fill out without many Dedicated Selection Pages, this is a good pattern to use. However, with the one or more Dedicated Selection Pages, this flow becomes a bit long.

Why Use It

Copying a desktop web form structure is a natural reaction for designers. Likewise, this form pattern is simple to use and understand for most customers that have used similar constructs on the web.

Other Uses

Any form can be considered a good candidate for this "default" pattern. Other form flow patterns in this chapter can be considered as optional improvements on this basic design.

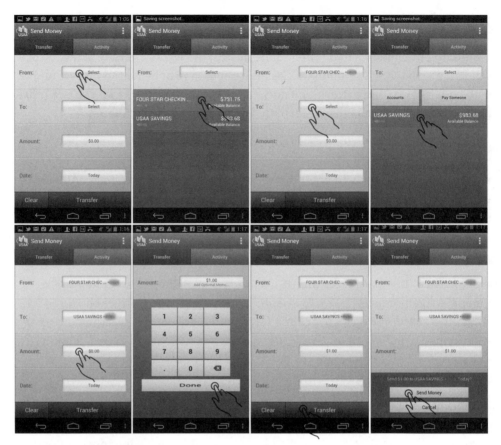

FIGURE 12.12: The USAA mobile banking app uses the Form First pattern for me-to-me transfers.

Pet Shop Application

In the Pet Shop Bank, you can use a simple variation of this pattern that is also going to take up the same number of pages—eight in total (see Figure 12.13).

The idea behind this pattern is simple: a copy of the standard desktop web form with dedicated selection pages used instead of the pickers and spinners. Using this mobile design pattern, the me-to-me transfer flow looks like this:

1. Blank form

2. Dedicated selection page for the From account

3. Back to the form (with the From account filled in)

4. Dedicated selection page for the To account

5. Back to the form (with both To and From accounts now filled in)

6. Fill in amount and other details and tap Continue

7. Verification page

Note the Verification pattern screen, which is recommended for use with the Form First pattern. In contrast with USAA, which uses the Toast Alert as a verification mechanism, the Pet Shop App uses a dedicated verification page. Read more about the Verification-Confirmation design pattern later in this chapter.

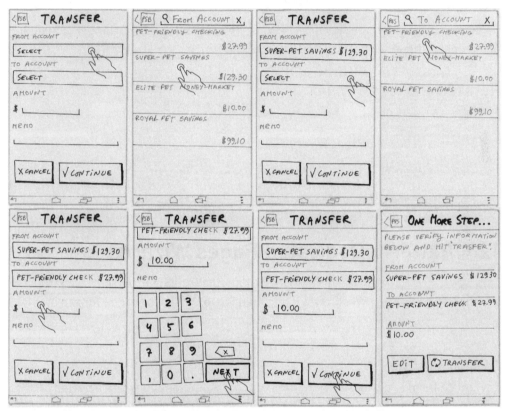

FIGURE 12.13: These wireframes show the Form First me-to-me transfer in the Pet Shop Bank app.

Tablet Apps

The Form First pattern is the primary pattern you should use for tablets as long as you don't take up the entire large screen for the Dedicated Selection Page. However, most apps don't take the time to create Dedicated Selection Pages in a pop-up or lightbox, creating long tablet flows where entire screens constantly slide in and out of view, using the entire large surface area of the tablet—frequent large-scale transitions within the same form can be distracting.

I've heard customers describe a typical Form First flow in a large tablet app for a large banking client as "underpowered," "childish," and "very long" mainly due to multiple large-scale transitions to and from the Dedicated Selection Pages. A better approach is to use the standard Drop Down control (covered in Chapter 10) for account selections or lightboxes for Dedicated Selection Pages as discussed in Chapter 14.

! Caution

Be aware of how long the flows are. If your flow is six screens or longer, consider using one of the other patterns in this chapter.

Related Patterns

10.5 Pattern: Drop Down

12.4 Pattern: Dedicated Pages Wizard Flow

12.5 Pattern: Wizard Flow with Form

12.6 Pattern: Verification-Confirmation

12.4 Pattern: Dedicated Pages Wizard Flow

This Dedicated Pages Wizard Flow can be called a "Mobile First" pattern (after a term coined by Luke Wroblewski) because it is entirely optimized for the small screen.

How It Works

The entire form is never shown to the customers. Instead they see a series of mobile-optimized pages with only the appropriate controls for data entry. Usually, at the end of the Dedicated Pages Wizard Flow pattern, the customer is presented with the verification page and a confirmation page.

Example

One great example of this pattern frequently overlooked is field extraction that happens automatically on the Android in the horizontal orientation of the device (see Figure 10.35 in Chapter 10, "Data Entry"). Thus, a long form consisting entirely of text fields in the vertical orientation turns into a succession of individual field entry pages optimized with the on-screen keyboard—a Dedicated Pages Wizard Flow.

Unfortunately, the pattern breaks quickly when it encounters a Select control in the middle of the form. Notice the last button in Figure 12.14 turns into Done, and the next screen shows the original form in the horizontal orientation. But why break the flow so abruptly? Would it not make sense to continue with the pattern until the entire form is completed?

FIGURE 12.14: The extraction of text fields turns them into Dedicated Pages, but this pattern breaks when it encounters a Select control in the Calendar app.

A better, more complete example of the Dedicated Pages Wizard Flow in a vertical orientation comes from the PayPal app running on the Apple iPhone (see Figure 12.15).

Note also the breadcrumb on the top of the page, which is a nice modification of this pattern that helps customers understand where they are in the flow.

FIGURE 12.15: The PayPal app's Send Money function on the Apple iPhone uses the Dedicated Pages Wizard Flow.

When and Where to Use It

Consider using this pattern whenever you have a short form with all fields required and where the fields utilize various specialized controls and customized keyboards, including one or more Dedicated Selection Pages.

Why Use It

The Dedicated Pages Wizard Flow pattern is ruthlessly optimized for mobile: Only the essential fields are included, which speeds up the data entry and guides the customer down the most efficient narrow selection path. This is a true "mobile first" way to implement a form and can be effective in keeping your customers in the flow. (Note: Discussion of flow is outside the scope of this book. For the seminal book on the topic, read *Flow: The Psychology of Optimal Experience* by Mihaly Csikszentmihalyi, 2008, Harper Perennial Modern Classics.)

Other Uses

You can use any short form with all fields required with this pattern. It also works if the field is optional, but be sure to provide a Skip functionality.

Some wizard flows are not dedicated—that is, the pages have two or more fields and do not automatically launch a keyboard. For example, the registration flow for creating a new Google ID that's shown in Figure 12.16 is not dedicated.

There is little evidence that breaking up the form this way provides any benefit whatsoever, but neither does it appear to hurt (see Luke Wroblewski's 2010 workshop *Mobile Input* at Design4Mobile in Chicago, http://static.lukew.com/MobileInput_LukeW.pdf).

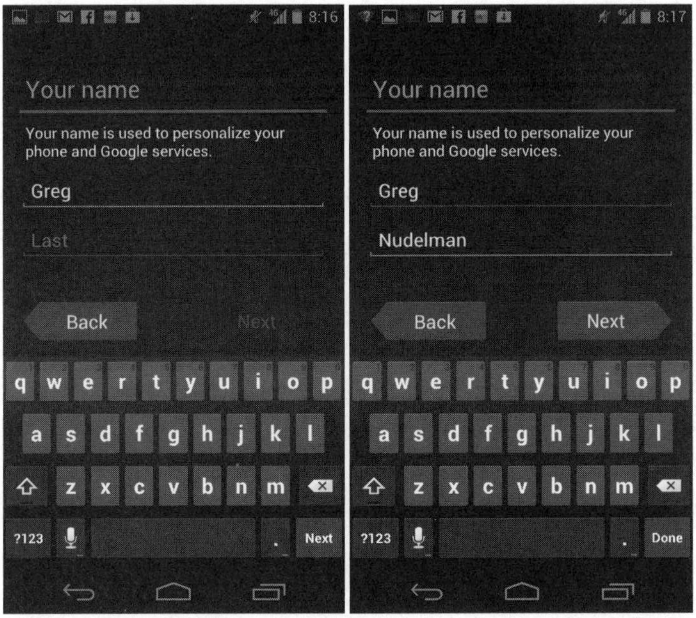

FIGURE 12.16: The form for creating a Google ID has a common (non-dedicated) wizard flow.

Pet Shop Application

For the Pet Shop Bank, you can imagine using this flow as a way to perform a fast me-to-me transfer with a minimum of fuss. The person selects the From account and the To account and then enters the amount using a dedicated screen with a numeric keyboard (see Figure 12.17).

The customer sees a verification page followed by a confirmation. The entire flow is accomplished in only *four* screens and the bare minimum of taps and keystrokes. This is a "steamroller express" pattern; to see why, contrast it with the much longer eight-screen Form First design pattern covered previously in this chapter.

Note that a Verification Page mobile design pattern (that enables the customer to review the entire transaction before tapping the final Transfer button) is recommended with this flow. Note also the use of the Breadcrumb mobile design pattern, showing the customer which step of the workflow she is on and how many steps remain. The Breadcrumb design pattern enhances the Dedicated Pages Wizard Flow pattern nicely.

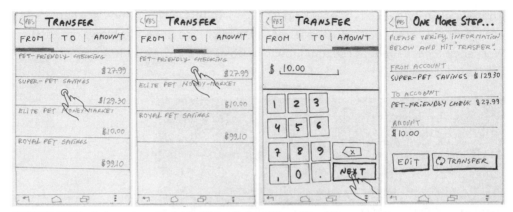

FIGURE 12.17: This wireframe shows the tightly efficient Dedicated Pages Wizard Flow pattern for a me-to-me transfer in the Pet Shop Bank app.

Tablet Apps

Tablet apps do not need this level of specificity unless the pattern is used entirely inside a lightbox.

! Caution

At first glance the Dedicated Pages Wizard Flow appears to be the sublime mobile perfection of a form that Man has been seeking since the days of the Greek statues. Should you create a Dedicated Pages Wizard Flow for every mobile banking flow on your site? Not so fast.

On mobile devices, nothing comes free. That includes the Dedicated Pages Wizard Flow mobile design pattern, which completely breaks down when used for longer forms. The primary idea behind this pattern is a dedicated page for each form element. If you have five or more form elements, the flow starts to become too long. Another issue is the inability of the Dedicated Pages Wizard Flow pattern to discern between optional elements (Memo) and required elements (Amount). Using this pattern, each element gets its own entry page with the appropriate keyboard and is likely to be perceived as "required." Even if the customer understands that she doesn't need to enter anything, each form element at least requires her to look at the additional page and click Continue to skip it.

So, is there another pattern you can use if you have five or more elements on the page and many optional fields? Glad you asked. One of the most versatile, yet under-used, patterns is the Wizard Flow with the Form pattern that's covered in the next section. And, as a bonus, this pattern dispenses with the need for a separate verification page.

Related Patterns

12.3 Pattern: Form First

12.5 Pattern: Wizard Flow with Form

12.5 Pattern: Wizard Flow with Form

This is the optimal mobile design pattern for use with the Dedicated Selection Page pattern.

How It Works

When the flow includes one or more Dedicated Selection Pages, they are presented first as a wizard. After the customer makes the choices needed for the Dedicated Selection Pages, he is presented with a form that contains the remaining elements.

Example

One excellent example of this pattern is the Chase Bill Pay flow mentioned previously in the Dedicated Selection Page pattern section. The system first presents the customer with a Dedicated Selection Page for selecting a Payee. The dedicated selection page is needed because of the complex functionality of adding payees on-the-fly that is also available. After the Payee is successfully selected, the system presents the customer with a form containing both required (Amount, From account) and optional elements (Memo) (see Figure 12.18).

The example in Figure 12.18 includes an additional verification page. That is not strictly necessary to use with this pattern because the final form acts as its own verification page of sorts.

When and Where to Use It

Any time your flow has one or more Dedicated Selection Pages, consider putting them first in the flow instead of the form, thus creating the Wizard Flow with Form pattern.

Why Use It

The reason for using this pattern is simple: It saves one tap and one page view compared to the Form First pattern for each Dedicated Selection Page. This pattern is the most flexible of all available approaches to filling out mobile forms.

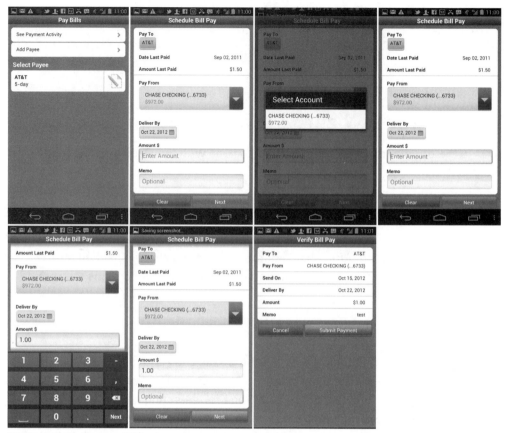

FIGURE 12.18: The Chase app Bill Pay feature uses the Wizard Flow with Form pattern.

This mobile design pattern combines the best features of the web forms, such as the flexibility of having optional fields and multiple input fields, with the vastly improved usability of mobile-optimized dedicated selection pages.

Another benefit of this pattern is the option to dispense completely with the verification page. With the Wizard Flow with Form pattern, the form page acts as a final editable verification page, dispensing with a need to have any additional pages. Of course, you can always add a verification page after the form if you must have it (refer to Figure 12.18).

Other Uses

The Wizard Flow with Form pattern also makes editing much easier than most other patterns. Instead of going through the entire flow again, the customer edits only the fields he needs. For example, in the Pet Shop Application section, to edit the To account, the customer taps the corresponding field in the form and the form navigates to the dedicated To account page. As soon as the customer selects the new To account value, the system immediately navigates back to the form without needing to go through the entire From account › To account › Amount flow again.

Pet Shop Application

The me-to-me transfer in the Pet Shop Bank is straightforward (see Figure 12.19) and adds only one extra tap versus the Dedicated Pages Wizard Flow (see the "12.4 Pattern: Dedicated Pages Wizard Flow" section).

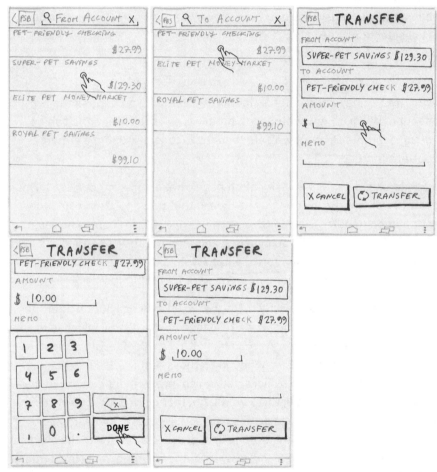

FIGURE 12.19: This wireframe for the me-to-me transfer uses the Wizard Flow with Form in the Pet Shop Bank app.

At the same time, this pattern allows for optional fields, such as Memo, to be shown. When compared with the Form First pattern, Wizard Flow with Form saves three taps and three intermittent views of the partially competed form screen. It also saves an additional tap because it dispenses with the need to have a Verification screen.

Because the form is seen only once and never disappears off screen, it makes its own effective verification screen, mostly dispensing with the need to have another.

Tablet Apps

This pattern is also a bit of overkill for most tablets, especially the larger ones. The crux of the pattern is the most efficient use of the Dedicated Selection Pages, which is rarely necessary on the tablet. Indeed, taking up the entire large tablet screen with the selection list is a bit unwieldy, unless you decide to use search or more complex functionality involved with the dedicated selection. So, for most Tablet forms, instead of using mobile patterns, consider using the popover/lightbox selection pattern to house the list of selections.

! Caution

Consider if you need to have dedicated selection screens. If you do, use the pattern in this section, Wizard Flow with Form; otherwise, if the selections can be performed using a simple Drop Down select control, consider using a simple Form First pattern instead.

If you do decide to skip a separate verification page, make sure you provide a Done button that removes from the screen any keyboard that pops up; doing so allows the customer an unobstructed view of the entire filled-out form before he taps the commit button. Actually, it's a bad practice and an antipattern to go to the next field if it's not required in any mobile form, and particularly in mobile banking where stakes are high and distractions must be kept to a minimum.

Unfortunately, the Chase app does exactly that—the Next keyboard button after the Amount field takes the customer to the Memo field (see Figure 12.20), which few banking customers would ever want or need to fill out. Yet this action of the Next keyboard button actually makes it look as though Memo is required. Worse yet, the keyboard displays the carriage-return character, which makes it difficult to exit the onscreen keyboard view (discussed in Chapter 10). Another drawback is that after the customer finally clicks off screen to dismiss the soft keyboard, the Memo field remains highlighted with a thick, bright, orange line, drawing attention to the one useless piece of information on the screen and away from the all-important Next button.

Instead, a better approach, as shown in the high-def mock-up in Figure 12.21, is for the Amount keyboard button to say Done and close the keyboard, so the customer would see the entire form screen.

If, at this point, the customer decides he wants to enter something in the Memo field, he can tap that field and enter the keyboard view again. The orange highlight is now in its proper place—around the Next button.

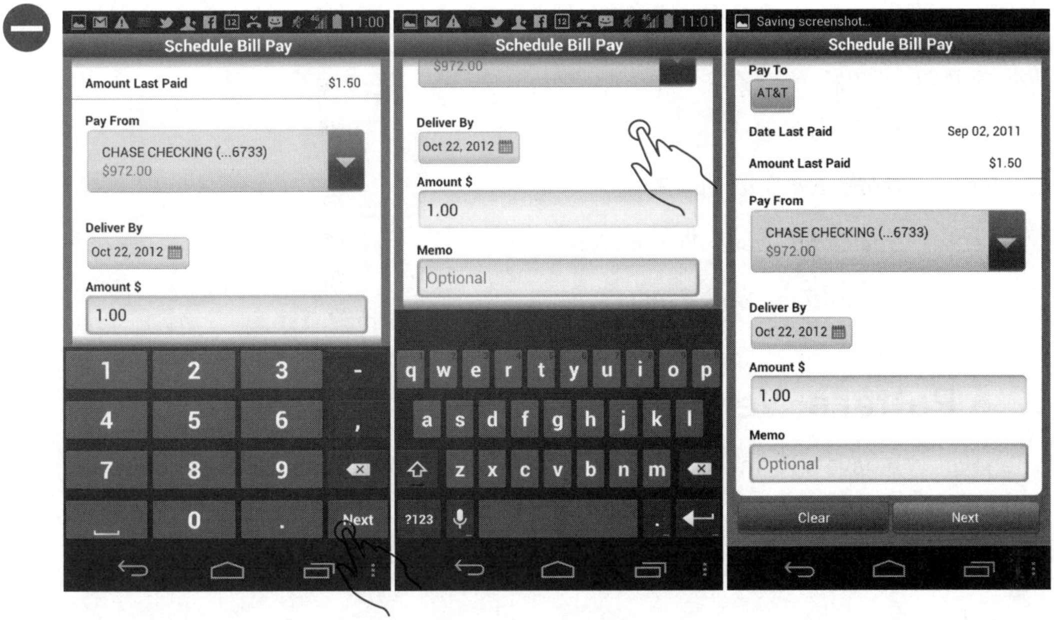

FIGURE 12.20: It's an antipattern to use a Next keyboard button that goes to a nonrequired Memo field in high-stakes flows.

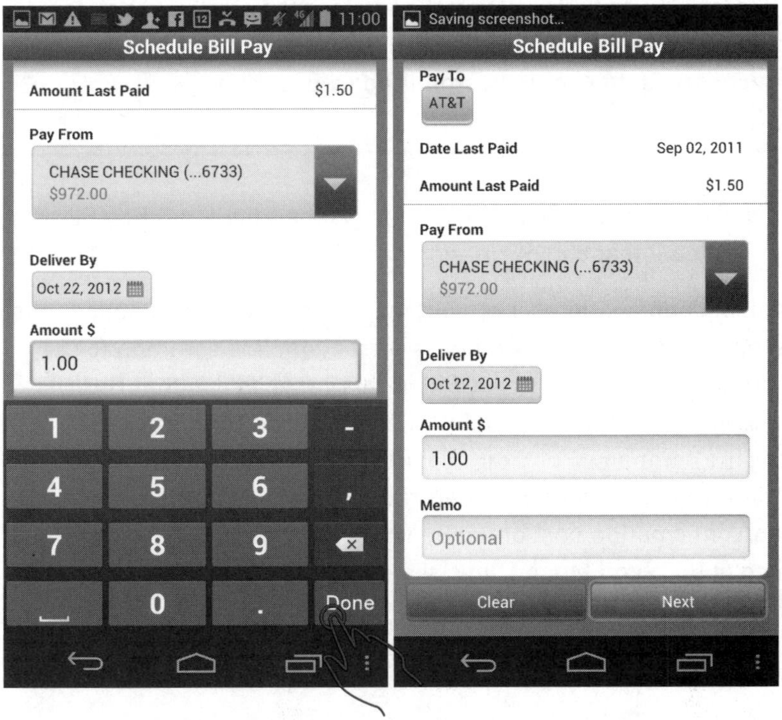

FIGURE 12.21: In this proposed fix for the flow in Figure 12.20, the Done keyboard button closes the keyboard and places the highlight on the Next button.

Related Patterns

10.5 Pattern: Drop Down

12.3 Pattern: Form First

12.2 Pattern: Dedicated Selection Page

12.4 Pattern: Dedicated Pages Wizard Flow

12.6 Pattern: Verification-Confirmation

12.6 Pattern: Verification-Confirmation

Whenever you have a form with a complex, emotional, or valuable transactional payload, you need to have a verification and confirmation of the action. This pattern deals with the best practices of this essential element.

How It Works

When the form is complete and ready for submission, the system presents one last screen for customers to verify the information before they tap the final commit button. After customers tap the commit button, the system presents the confirmation page, showing the details of the committed transaction.

Example

One example of the Verification-Confirmation pattern is the Chase Bill Pay Flow, which presents the verification and confirmation pages in turn (see Figure 12.22).

The buttons on the final confirmation screen outline the most common actions the customer might take immediately following the transaction. It is common to see the primary next action button (Pay Another Bill in this case) as one that returns the customer to the beginning of the original flow, especially if this happens to be the "money flow" for the company. For example, PayPal makes its commission every time you send money to someone; so naturally, it's taken the time to optimize the verification/confirmation part of the Send Money flow to put the customer back at the beginning of the Send Money flow (see Figure 12.23) to ostensibly send more of their hard-earned cash to someone else. Note that the PayPal app's confirmation page is implemented as a lightbox for just this purpose.

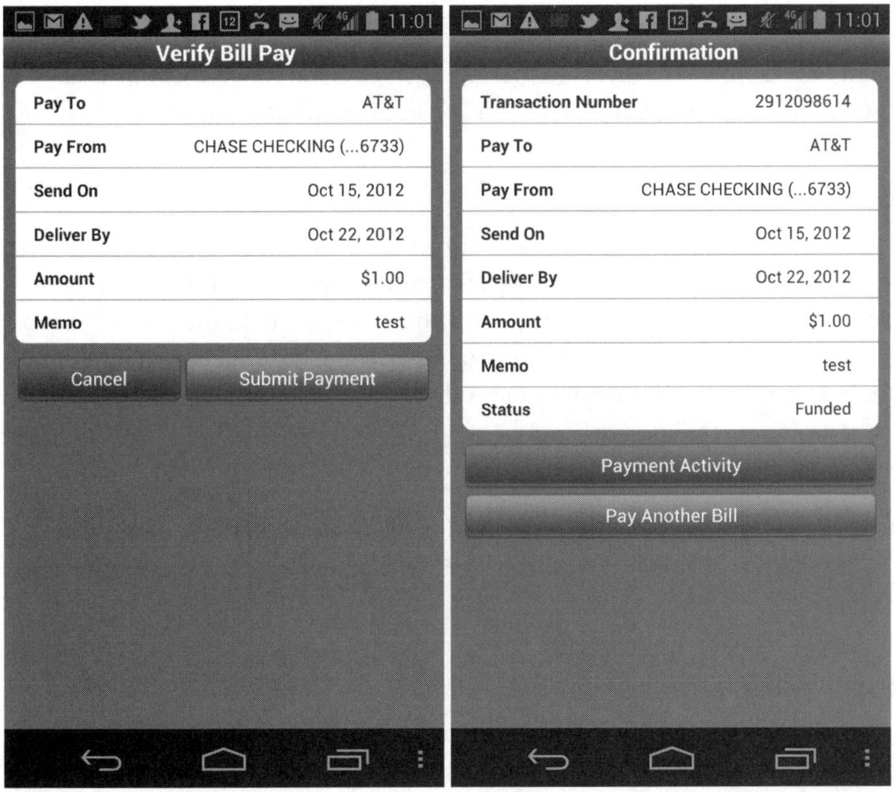

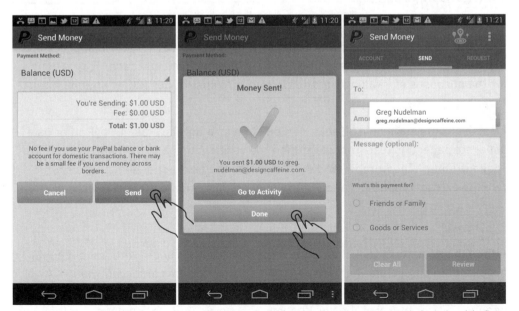

FIGURE 12.22: This is how the Verification-Confirmation pattern looks in the Chase app Bill Pay flow.

FIGURE 12.23: The confirmation implemented as a lightbox in the PayPal app returns the customer to the beginning of the flow.

When and Where to Use It

Any time you have a flow that involves money, such as banking or e-commerce checkout, customers expect some sort of a Verification-Confirmation pattern.

Why Use It

Verification enables the customer to make sure she did not miss anything or accidentally mistype something in the fat-fingered mobile apps world. This is especially important due to frequent interruptions and auto-corrects that might creep up during data entry. The extra verification page provides visual confirmation that the system interpreted the data entry correctly.

Other Uses

Sometimes, you can dispense with the separate verification page for short forms. See the "12.5 Pattern: Wizard Flow with Form" section earlier in this chapter for more information.

Pet Shop Application

You've seen the verification page in most of the designs in this chapter. It's a straightforward design, so it's not necessary to replicate it here.

Tablet Apps

For tablets, verification on a separate dedicated screen may seem like overkill. However, sometimes, especially for large financial transactions, it is appropriate to dedicate the entire separate page just for the verification screen. This is done for the purposes of "waking up" the customer to ensure her full attention on the task.

Effective confirmation on a tablet can be done as a lightbox (as in PayPal) or as a separate page. It is less important than the verification page because by then the "deed is done." The number and desirability of the next actions in large part dictate the extract implementation of the final confirmation page.

! Caution

Although this pattern looks trivial, it is far from that. To begin with, most people have trouble differentiating between the words *verification* and *confirmation*. As a result, unless your design makes absolutely clear that the transaction is not done, people may misinterpret the verification page as a confirmation page. In other words, upon seeing the verification page they think that they are "done" and do not press the final commit button (which might be something like Place Order, Submit Payment, or Transfer) and exit the flow instead.

Almost 30 percent of the 25 participants in one massive usability test for a major Internet retailer made the same mistake in a long checkout flow. When these people saw the verification page, the design of the page suggested to them that they were done. This, as you can imagine, is a dangerous and nasty usability problem. Think about it: The customers determined to hand over their money, instead stopped short due to confusion—nothing can be more unfortunate for the retailer. And think about the customer frustration when they wait for weeks for goods that never arrive!

The bad news is that this problem can be difficult to diagnose. Look for unusually high percentages of people abandoning the shopping cart or unfinished transactions. One of my clients, a major U.S. bank, saw transaction abandonment in the 20 percent range, which is acceptable (though highly unwanted) for sending someone money, but it's too high for me-to-me transfers. Seeing unrelated transactions like this that have similar abandonment rates is a good clue that there's something going on other than people simply changing their minds. If you are at all in doubt, run qualitative usability tests on your system. Don't forget to check in with your customer support. Are people calling in to ask, "Where is my shipment?"—that's a good clue you screwed up your verification page.

One of the most common mistakes in this pattern is placing the final commit button below the fold. The combination of the final commit button below the fold and the bad copy on the top of the page (such as a page title of Verification) is a real conversion killer. The solution is to place the final commit button both on the top and the bottom of the page and name the page with a "transitional" name such as Review Your Order or Checkout. Amazon.com is an excellent example of an effective Verification-Confirmation pattern (see Figure 12.24).

You can give your verification pages even stronger "incomplete" titles, equipped with an ellipsis, such as "One More Step..." and "Are you Sure?" Make your final commit button large and bright with a highly contrasting color. For example, Amazon.com's, final commit button is a highly saturated yellow.

Related Patterns

All patterns in this chapter

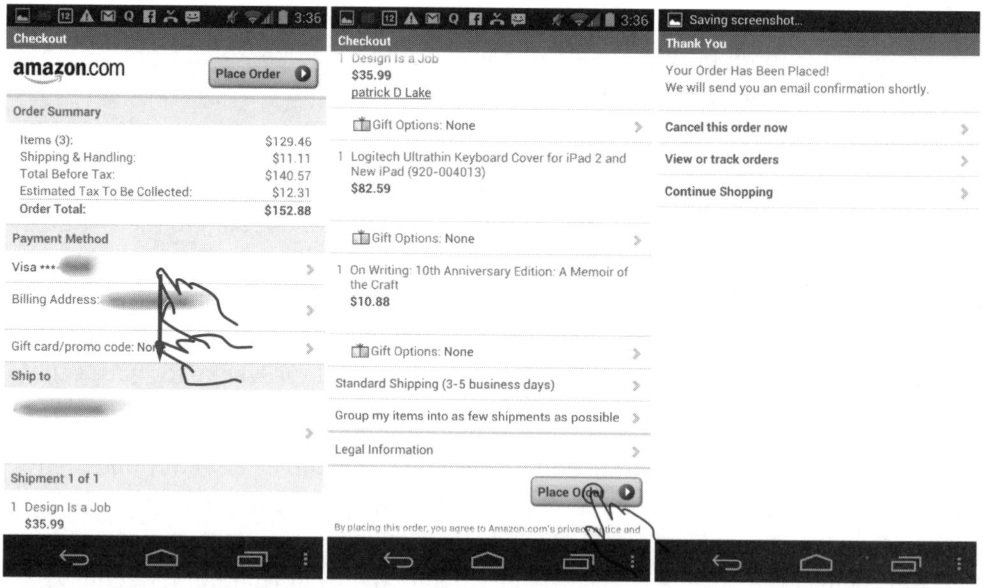

FIGURE 12.24: The Amazon.com app's checkout process is an excellent example of the Verification-Confirmation pattern.

12.7 Pattern: Near Field Communication (NFC)

Telecoms and mobile phone manufacturers have long been promising mass-market mobile near field communication (NFC). With a brand-new Galaxy Nexus S, the first NFC-enabled smartphone on the market, consumer mobile NFC has suddenly become the new reality.

How It Works

NFC is a short-range connectivity technology. Connecting with NFC usually initiates apps or other systems within the device. NFC enables easy access to these applications by touching two NFC-enabled devices to one another or touching an NFC-enabled device to an NFC "tag."

Example

One shining example of the use of NFC in the U.S. market is Google Mobile Wallet. When it's time to pay, the customer unlocks the wallet using a four-digit PIN and taps the mobile device to the pay terminal. The timer runs until the payment information is captured by the terminal (see Figure 12.25).

FIGURE 12.25: Google Mobile Wallet uses the NFC pattern.

Despite the apparent simplicity, there are many issues with the current NFC flow of the Google Wallet app:

- Security of the wallet app

- Customers' understanding of how the app works

- How to show which card is being used.

Security of the Wallet App

With the new doorway into the phone's innards, *security* is likely to be paramount. How is the access to the NFC chip's impressive capabilities maintained? If you use the app for NFC-enabled Foursquare access, you may not even need a password. For Google Wallet, the app uses a decidedly iPhone-esque four-digit PIN screen, as shown in Figure 12.26.

I guess the thinking behind this design at Google HQ was that if a four-digit code can be used to unlock the ATM machine so the customer can get $300 at a time, this same security ought to be sufficient for a phone wallet. This may be the case, but if you dig a bit deeper, you discover a host of important questions.

FIGURE 12.26: NFC Google Mobile Wallet maintains security through a four-digit login PIN.

For example, when the NFC-enabled app is launched, how long does it stay active? The ATM PIN is only active while the transaction is occurring. If the customer pauses for more than 20 or 30 seconds, the standard-issue Diebold ATM boots the customer from the system. Although an ATM can be considered to be a crude version of "mobile technology," strictly speaking the ATM machine can virtually guarantee your physical presence at a particular location (in front of a specific ATM). The ATM can also guarantee to command your complete attention. You cannot, for example, check your Facebook alerts while also taking the money out of the ATM.

In contrast to an ATM, on the smartphone you are dealing with the infinitely more complex device and multivaried mobile environment. For example, on the mobile phone, the customer may launch the app in preparation for payment while waiting in line. This is tantamount to someone entering his PIN while waiting in line for an ATM. Should you allow it?

If you do allow it, how long do you allow the app to stay active? Five minutes? Four? Or until the phone is shut down? How about multitasking? Should you allow the customer to check his e-mail while his digital wallet is "hanging open"?

How about the four-digit PIN? Most consumers might use for their digital wallet the same four numbers they use to withdraw money out of the ATM. That seems logical, if somewhat simple-minded.

How about using the same 4-digit PIN for the mobile wallet and the ATM as they use to unlock their phone? Now that seems like a decidedly terrible idea because anyone getting access to the phone now also has access to the digital wallet and the ATM. Yet most people will not hesitate to use the same code because remembering several separate, but not identical four-digit codes and remembering when to use each one, is a chore.

Customers' Understanding of How the App Works

Wave here! Tap there! Why is it that the simplest technology often seems the most complicated? It's easy to miss the simple step of explaining how things work to the customer when it seems so obvious. Take a look at the Google Wallet home screen in Figure 12.25, for example. A naïve customer might ask, "How do I use this?" From the engineering standpoint for which Google is so famous, this seems obvious—just wave the dang thing over the payment terminal already!

However, from the customers' standpoint, the last thing they want is to look like a fool with their stupid, shiny, new NFC phone, "that doesn't even work," while they fumble for "real money" in front of their friends or significant other.

Just search for Google Wallet on YouTube and watch some of the videos. People are unsure. This NFC thing is new, and people don't know what to do with it. Until a new technology enters the consumer consciousness, you need an *Inukshuk*, which is an Inuit User Experience term coined by the incomparable Jared Spool to indicate content that provides little factual information. The sole purpose of Inukshuk content is to cradle the customer in the luxury of care and comfort. Inukshuk is a little human touch in the information-dense digital universe. "Someone has been this way. They have used this. It works. You won't look like an idiot. You won't lose money. It's alright. Let me show you."

How much better a newcomer's experience would be if Google took care to include just one tiny Inukshuk. One additional button—How to Pay—as shown in my wireframe in the later "Pet Shop Application" section can make a difference. Fortunately, the NFC functionality is currently limited, which gives you plenty of real estate, even on the tiny mobile screen.

A How to Pay section would also be the perfect place to educate the consumer about security, timeout, and other useful considerations. It would also be a great place to introduce the inevitable complete money movement functionality, such as Bill Pay, Person-to-Person payments, and Inter-Account (also called me-to-me) transfers. This Inukshuk can always move to the Settings screen after the customer reads the content and accomplishes a few successful scans, or could be turned off with a Don't Show Again switch.

How to Show Which Card Is Being Used

Lastly, there is an important question of default identities. Today, most of us have multiple online identities. In my case, I have a small business. Sometimes I act on my own, as Greg Nudelman, and sometimes I act as a corporate entity and the CEO of my company, DesignCaffeine, Inc.

Unless you can be sure that your customers will be carrying two NFC-enabled phones, each with a different digital "identity" and a different digital wallet, they will have a need to determine, quickly and with a high degree of precision, which of their identities is currently selected on the device (and which of the credit cards will be presented to the NFC reader).

Unfortunately, Google Wallet does not make that easy. Can you tell which credit card has been selected in the home screen referred to in Figure 12.25? Actually, you cannot tell from only the home screen. You must drill down into the Payment Cards screen, as shown in Figure 12.27.

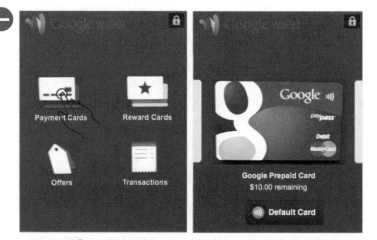

FIGURE 12.27: To see which credit card is selected for payment, a Google Wallet customer needs to drill down to the Payment Cards screen.

NFC makes it easy (almost too easy!) to send the right message but as a wrong person. Thus showing the default identity setting, user ID, credit card, and so on is important. Fortunately, a carousel navigation pattern like the one shown in the Pet Shop wireframe in Figure 12.29 would make a default credit card easy to see and change.

When and Where to Use It

Most of the apps currently using NFC are just "testing the water" because the NFC technology is still young. Consider using NFC any time your customers would benefit from connecting their physical and virtual worlds, such as for social media check-ins, point-of-sale m-commerce (mobile commerce), getting more information, and meta data about many kinds of physical objects.

Why Use It

NFC has the power to connect mobile devices to every manufactured object and landmark, thereby forever binding the physical and virtual worlds and delivering on the promise of Bruce Sterling's "Internet of Things" (*Shaping Things*, 2005, The MIT Press)—readily available meta data, social media data, expiration date, manufacturing history, and much more about every object in our environment. Besides replacing completely QR codes, banking cards, ATMs, advertising, identity cards, and cash as you know it, there are many other uses of NFC that have not even been invented yet.

Other Uses

Consider all your NFC-enabled functions as "nice to have" but also implement a robust and tried alternative in case the customer who wants to interact with your NFC smart tag did not run out and purchase a Galaxy Nexus S yesterday. Make sure you take care of your existing customers while making things a bit smoother for new folks.

For example, say your app can use NFC for checking in at a location. Consider that many of your customers may not yet have NFC-enabled phones. Should you still provide the service to these folks? If you want more than six people to show up, the answer is a very emphatic Yes.

You can marry the old and the new by combining the tried and true QR Code with an embedded smart NFC tag. To let customers know that there are two ways to interact with the tag, start by embedding the standard NFC "wave" symbol into the QR code to indicate dual functionality. See the example in Figure 12.28.

This tag design makes it clear that NFC-enabled devices can simply tap the QR code sticker, whereas those who carry Android phones without the NFC chip can scan the tag with the QR Code Reader. This way all of your potential customers can take full advantage of your service. Needless to say this approach may not work for mobile banking (unless it's a one-time QR code generated on the screen during the checkout); it is presented here as merely another application of the NFC technology.

FIGURE 12.28: The NFC-enabled QR Code tag combines the QR Code with NFC technology.

Pet Shop Application

As discussed earlier, two important improvements are added to the design of the Google Wallet: a "How to Pay" button and a carousel navigation pattern to make it easy to see and change the default credit card. The resulting Pet Shop Bank Wallet design is shown in the hi-def wireframe in Figure 12.29.

FIGURE 12.29: The Pet Shop Bank app Mobile Wallet solves some current UX issues.

In this version, swiping across the cards at the top changes the card to the next one in the wallet; and double-tapping any card brings up the entire card list with thumbnails. In addition, a tap-and-hold brings up a menu that enables customers to see the list or add a new card. Implementing double-tap and tap-and-hold should make this virtually fool-proof while not interfering with the primary action of swiping across to get to the next card. The carousel is meant to be similar to viewing the actual stack of cards inside a physical wallet. Of course, this needs to be tested in the field to ensure that virtual foolproofness.

You might say that that I've taken a beautiful, clean home screen designed by Google and made it a mess. And you would be right. The biggest take-away here is that until some of these questions are figured out, NFC can be quite messy where user experience is concerned.

Tablet Apps

Don't forget tablets, especially the smaller ones. Currently, there are no tablet devices on the market capable of NFC interaction out-of-the-box. However, NFC is definitely going to be a near-future tablet development. That said, I imagine little difference in interaction with an NFC terminal between a 7-inch tablet and a mobile device. All the same considerations discussed here still apply.

! Caution

NFC on mobile devices is still fairly nascent technology. Here are the key points to consider as you design your own NFC-enabled app.

Design for Evildoers

Don't forget to carefully design the experience that the evildoer that stole your customer's phone and is actually trying to break into the NFC app will have to endure. Should the app lock up after three tries? Ten tries? Should you give the customer a progressively longer cooling-off period after three unsuccessful tries, iPhone-style?

These are not straightforward questions, and the answer just might be, "It depends." If the app is used to get onboard a commuter train, perhaps a 2-minute timeout with 10 tries to lock up might be the right answer. But if the app happens to be a digital wallet, 2 minutes for a timeout and 10 tries for a lock-up might be a bit long.

Security, PINs, and timeouts all bear careful consideration in the design of your ideal encounter of the NFC kind. Some field-testing is certainly necessary to pinpoint possible issues.

Is This Thing On?

NFC functionality is always on by default. Few people give any consideration to this. Like the cell phone signal, NFC is continually transmitting and receiving even when the phone is "dark"—that is on standby. Should this be the case, or should the customer have some control over where and how NFC is presented? Should the NFC tag determine which app to call when the NFC signal is obtained, or should the customer have some control over which app is launched? And should the mobile device confirm with a customer before launching that app, and what it will do? Or will it allow the customer to profess to the entire world his new adoration of Bed Bath & Beyond on Facebook with a single foolhardy tap? How does a consumer (or, more properly, an actor in the NFC service) smoothly and naturally determine the extent of the role he wants to play without going all the way to the advanced settings in the Beyond department?

It's only a matter of time until complete strangers will try to "bump" or "wave" your customers' phones, *often without their permission*. If the goal is to avoid dropping the NFC down into the same technological dead end to which previous near-field attempts such as Bluetooth have been relegated, designers and developers must give careful consideration to the question, "Is it on?" Does the app need to be active to interact with the external NFC tag? What if the app is active but runs in the background in the multitasking mode? Does the phone need to be turned on, or can it be on standby? What information, if any, can be requested by the NFC terminal? How does a customer control what will be sent and to whom? What can be obtained from the customer's phone without his permission? How hard or easy it is to connect to something that carries the NFC signal?

Sometimes, the rapid pace of mobile technology adoption reveals more questions than answers. But that's exactly the mystery that makes Android mobile and tablet UX design so intriguing.

Related Patterns

12.1 Pattern: Login Accelerator.

CHAPTER 13

Navigation

Some of the navigation patterns have "bled" into previous chapters of the book and for good reason: Navigation is the third most important topic of the book. Together with search and data entry, navigation forms the triumvirate of effective mobile User Experience. And Android navigation is a huge topic—one that deserves its own book. Unfortunately, it has only one chapter, so it includes only the most advanced, hotly debated, and most commonly screwed up topics as subjects to provide the most value to you.

This chapter explores the patterns that make navigation possible and even enjoyable on tiny mobile screens. Topics such as panorama integration and immersive experiences might be just around the corner, but they have not yet found a way into the collective Android design consciousness. As new, specific mobile navigation patterns continue to evolve on a daily basis (almost), it becomes more important to focus on the solid, authentically mobile design foundation that enables you to adopt new and emerging mobile patterns to your needs.

⊖ 13.1 Antipattern: Pogosticking

What sounds like a fun childhood game becomes an extreme sport on the small bit of real estate offered by mobile screens. Here is how to avoid breaking a leg.

When and Where It Shows Up

Pogosticking is an insidious mobile navigation problem. It shows up any time there is a list view linked to a more detailed view that contains additional information.

Example

Pogosticking is frequently an issue for the larger desktop web screens, but it is even more of a problem for mobile. For example, the popular app TripAdvisor displays only a fraction of information about a hotel in the list view: only four pieces of information (see Figure 13.1). This makes navigating such a view a challenge.

⊖

Tamas Estates

#**11** of 13 *attractions in Livermore*

Winery, Shop

Call Us

Get Directions
8.99 mi E

FIGURE 13.1: This example of the Pogosticking antipattern is from the TripAdvisor app.

The small size of mobile screens does not automatically guarantee that you must skimp on list-level item information. Compare the TripAdvisor app (refer to Figure 13.1) with the Yelp app list view, as shown in Figure 13.2. Yelp shows customers six pieces of information—that's two more than TripAdvisor!

2. De La Torre's Trattoria   0.9 miles
6025 W Las Positas Blvd, Pleasanton $$
★★★★☆ 92 Reviews   Italian

FIGURE 13.2: This is Yelp's solution to the Pogosticking antipattern.

Moreover, Yelp manages to alleviate the pogosticking issue while actually showing more search results on the same size screen—approximately 1.5 Yelp results for each result shown by TripAdvisor, which is quite an accomplishment. Yelp does this through improved layout of the individual list result.

Why Avoid It

As UX expert Jared Spool—who coined the term *pogosticking*—explains, most of the effective navigation decisions are made on the gallery (list) pages, and drilling down should be necessary only to actually engage with the item. In pages with excessive pogosticking, the gallery or list page does not provide enough information to make a good navigation decision, so the customer must jump in and out of the multiple detail pages like a child on a pogostick. This wastes valuable time and customer attention span and detracts from productive activities that make your company money, connect, inform, and entertain, which are the activities you want your customers engaged in. Worse yet, due to excessive pogosticking, customers may never find what they are looking for and give up trying, abandoning your app altogether.

Additional Considerations

Pogosticking is a big enough topic to warrant its own chapter, which you can find in my book *Designing Search* (Wiley, 2011, ISBN: 978-0-470-94223-9).

Related Patterns

All patterns in Chapter 7, "Search."

⊖ 13.2 Antipattern: Multiple Featured Areas

All customers enjoy a good deal. And if one type of featured content is good, more *must* be better, right? Wrong. The Multiple Featured Areas antipattern is one of the most common mistakes in the mobile space.

When and Where It Shows Up

This antipattern shows up whenever there is more than one type of featured results. This is especially common when marketing folks step in, insisting on different brand-named featured areas with funky flavors, usually to match the website.

Example

Examples of this antipattern abound everywhere you look. One of my favorites is in the NewEgg app, whose homepage proudly features Shell Shocker deals, automatically rotating Daily Deals and the EGGXTRA! EGGXTRA! section (see Figure 13.3)—all in addition to "normal" search results (that is, if people can figure how to get to them).

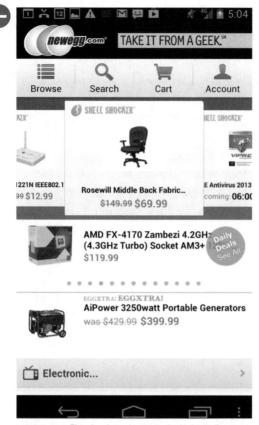

FIGURE 13.3: There's a lot going on in the Multiple Featured Areas antipattern in the NewEgg app.

Why Avoid It

Prominent and respected UX practitioners like Steve Krug and Jacob Nielsen have reported many times that branded controls and offerings confuse customers. This is completely supported by my experience doing research for top e-commerce firms. That goes triple for multiple branded offerings on a small mobile screen. What's the difference between the different types of deals? *Who cares?* Most of your customers certainly do not, and on a smaller screen, over-promotion quickly turns into missed revenue opportunities.

If your customers can't explain to you clearly the difference between these various marketing offerings, chances are they are instead confused, asking themselves, "Am I actually getting the best deal with a Shell Shocker? Or should I look at the EGGXTRA! EGGXTRA! item instead? What if the best deal on a new tablet can be found under the Daily Deals?" Now instead of actually *buying* something, the customer is flipping between the marketing promos trying to figure out how to make sense of your bad marketing plan. In the UX world, the term for this behavior is *churning*—unproductively navigating between useless entries that do not meet the customer's goals.

Additional Considerations

If you get tired of fighting the marketing folks and want to work *with* them by offering options that support the free expression of their awesome marketing genius without the negative impact on the mobile customer experience, consider the Amazon.com app model: different deal types aggregated under the same title of Gold Box (see Figure 13.4).

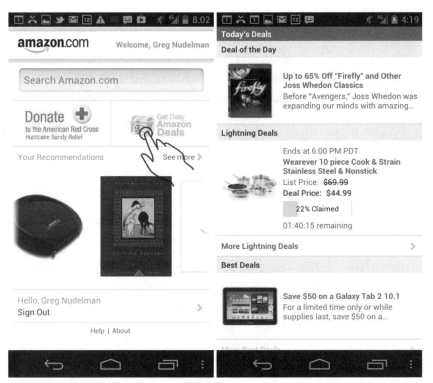

FIGURE 13.4: The Multiple Featured Areas antipattern in the Amazon.com app is hidden under Gold Box.

Yes, these are all different deal mechanisms, names, and branding. What makes this example different from the NewEgg Shell Shocker mentioned earlier is that Amazon.com did not attempt to show all the featured items on the homepage. Instead, the promos are generally well hidden under the Gold Box label, which is somewhat confusing and ambiguous, so most customers are not assaulted by the bevy of confusing options unless they deliberately aim to seek them out. In which case, the over-the-top branding of the Gold Box is a fair warning for the uninitiated that they are about to step into an unknown and possibly confusing territory of discounted, featured products.

Related Patterns

See all patterns in this chapter for alternative navigation ideas.

13.3 Pattern: Carousel

Speaking of browsing featured products, one of the best navigation patterns for a small collection of visually appealing products is the Carousel pattern.

How It Works

The customer sees several images of products across the row. To explore more products the customer can swipe across the row to navigate horizontally to the next set of products. An arrow indicating the direction of the Carousel movement is usually provided as a clue for the required interaction. Alternatively, one of the products may be partially hidden, creating a *teaser*, indicating that more content can be visible by swiping.

Example

One excellent example of this pattern is the Amazon.com app's home screen (see Figure 13.5).

This implementation of the Carousel pattern uses the teaser method of suggesting the required interaction. In Figure 13.5, only a small, tantalizing glimpse of the naked CAT5E Ethernet cable is visible, which is completely irresistible to its more impressionable customers, who can't *help* but swipe across to get to more content.

When and Where to Use It

You can use this pattern any time you have a small set of 8 to 20 products or items that are easily recognizable by their picture.

Why Use It

Carousel is an attractive and still fairly unusual control for presenting visual information. Carousel takes full advantage of the multitouch gesture of swiping that is available on the mobile device. Carousel is easy and intuitive to operate and takes full advantage of the compressed mobile real estate when few words are needed to support the content.

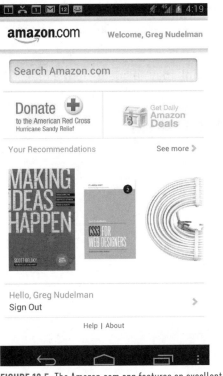

FIGURE 13.5: The Amazon.com app features an excellent example of the Carousel pattern on the home screen.

Other Uses

One of the best features of carousels is that they work well in a wide variety of device sizes and screen resolutions. This includes the ever-tricky horizontal orientation (see Figure 13.6), where Carousel works even better than it does in the vertical orientation.

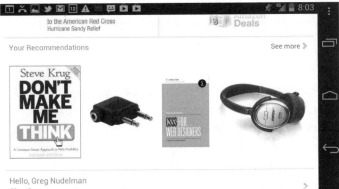

FIGURE 13.6: The Carousel pattern in the Amazon.com app adjusts to various screen sizes and works even better in the horizontal orientation.

Whereas the usefulness of the traditional search results would be severely affected by the lack of vertical space, a well-designed Carousel shines by showing off even *more* inventory.

Also noteworthy is the presence of the See More link that navigates to the featured search results (see Figure 13.7).

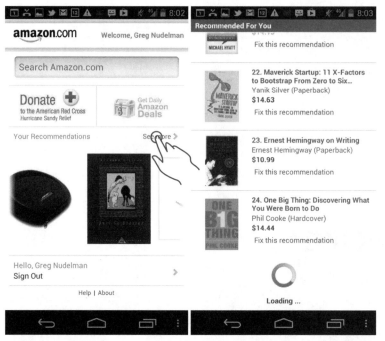

FIGURE 13.7: The See More link in the Carousel pattern navigates to the featured search results.

This is an excellent idea if your small subset of items in the carousel control fails to meet the customers' needs but piques their interest. You can find more on this pattern in the 2-D More Like This pattern in Chapter 14, "Tablet Patterns." In this instance, the entire carousel control acts as an advertisement of sorts, engaging the customer in exploring the relevant corner of the massive Amazon.com inventory.

Pet Shop Application

The Pet Shop app can use the carousel to show off featured pets or relevant new arrivals. For example, new items matching the customers' last search of "Guard Dog Puppies" in their local area would be a sure winner. See the wireframe for the 2-D More Like This pattern in Chapter 14. (Looking for the source code for this pattern? Head to http://androiddesignbook.com for the downloadable mini demo app of the Carousel pattern complete with source code you can use right away in your own app.)

Tablet Apps

This is an exceptionally great pattern tailor-made for large swiping gestures that tablet devices invite. The Carousel pattern on tablet devices is even better as a 2-D set of carousel controls. See the 2-D More Like This pattern in Chapter 14.

! Caution

With any pattern, there are many ways to create an implementation that does not feel quite right. One example is the NewEgg app's Shell Shocker Carousel implementation, as shown in Figure 13.8.

FIGURE 13.8: The NewEgg Carousel has a few issues.

The discussion of UX issues with the NewEgg implementation would prove instructive. When designing a carousel control it helps to think of its real-world counterpart: the carousel ride at the country fair. Here are a few recommendations to make your carousel ride the best of the bunch.

- **Scroll smoothly:** To begin with, the NewEgg carousel is structured more like the Apple OS cover flow, with the large central element and two partial views on each side on the periphery. Like the Amazon.com carousel, this one can be swiped to advance through the list of additional products. However, the New-Egg carousel moves jerkily because of the structure of the central element, making it hard to see the intermediate states during scrolling, which is a major disadvantage. It is particularly hard to see the two peripheral elements changing; things are hopping all over the place, instead of smoothly swimming by the way they do in the Amazon.com app. Higher-end and classic carousel rides ease in and out of the speed smoothly and provide a pleasant, mellow, smooth ride. A Carousel is a high-end visual viewing experience that is meant to promote flow, not stress out the viewer. All parts of the control, including transitions, need to work smoothly together.

- **Indicate initial scroll direction:** NewEgg appears to be scrollable to both the left and right directions, creating confusion: Is this Carousel meant to represent a circle? Have you seen everything already or do you need to keep scrolling? Amazon.com uses the standard Android 4.0 screen tilt boundary treatment to signal the end of the line—a much better approach (if the customer attempts to scroll past the end of the carousel, the screen content tilts to indicate that further navigation in this direction is not possible). Just like a real carousel has horses that point in a certain unambiguous direction (you would not sit backward on a horse, would you?) so your own carousel implementation must show which way the ride runs.

- **End the ride quickly:** The better carousel implementations indicate the end of the list with the same screen tilt boundary treatment as the beginning, and only present 8 to 20 items, after which the ride is over and the customer can get off the carousel. The customer needs to leave the ride with the feeling of *still wanting more*. In contrast, the NewEgg carousel seems to go on forever, so the customers do not get off until they are feeling bored (or given the jerky transitions, more likely weak in the stomach). A much better approach is to use the last element in the carousel to provide an obvious built-in More Like This link for jumping into the search results that are more efficient for scanning through large quantities of data and are more likely to be relevant through sheer volume of items (see the 2-D More Like This pattern in Chapter 14 for an example of how this looks).

- **Make sure your horses look amazing:** No matter how smooth the ride is or how far or how fast the carousel goes, the best carousels have the best-looking horses—period. Make sure your picture thumbnails (the horses) tell the story you want your audience to see. For example, the Amazon.com thumbnails are much better looking than those from NewEgg; although sometimes even the m-commerce giant screws up the ride, dropping the thumbnails entirely (see Figure 13.9).

It goes without saying that ghost horses make for a terrible ride, even on Hallow-een. Which brings up the next point: Some products are just not that visual, which makes them poor candidates for inclusion in the Carousel. It is sometimes better to have some text in the individual tile to provide an additional information scent for the technical gadgets, like those sold by NewEgg. If the picture tells only half the story and you have to include a great deal of text, you have to increase the size of the individual element to the point where the carousel may no longer be the best presentation. For those items think twice if you even need a carousel, or if a simple vertical list (more akin to a themed rollercoaster ride) would create a bet-ter experience.

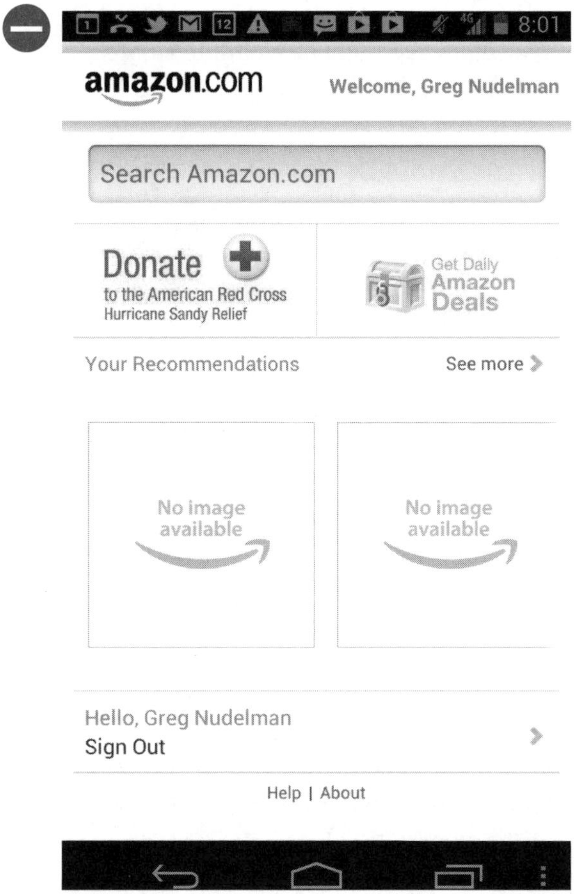

FIGURE 13.9: Amazon.com's Carousel sometimes omits the thumbnails.

Related Patterns

14.5 Pattern: 2-D More Like This

13.4 Pattern: Popover Menu

On the desktop, you have the right mouse button to provide a contextual menu. However, the desktop web historically has had access to only a single mouse button, so it introduced the concept of an action menu activated by a left-click.

The mobile platform is special because the left and right mouse buttons do not immediately translate to touch. However, the first iPad version of the iOS introduced the tap-and-hold popover menu that provides additional menu functionality similar to that of the mouse right-click (tap-and-hold is also common on Windows Phone for contextual/secondary actions and information). Although tap-and-hold menus are a creative and interesting solution, they have many discoverability issues. In the Android platform, the emerging dominant paradigm seems to be a simple single tap menu that expands the choice of actions available to the customer.

How It Works

When the customer taps the "actions" element or arrow, the popover menu of actions opens to reveal more choices in a popover element on top of the existing content.

Example

One interesting example of using the popover menu for navigation comes from the LinkedIn app. Tapping the logo opens the navigation layer as indicated by the down arrow under the logo (see Figure 13.10).

Another example of the similar popover used under the menu is one from the Wapedia app, where the popover menu is used mainly to attenuate the search domain. (See Figure 13.11.)

This is *siloed search*, the inverse of the usual pattern where a single search is performed over various databases (what is known as *federated search*, such as Google or the Library of Congress.) In the Wapedia example, the customer must manually switch between various databases, providing what Peter Morville calls "advanced, focused queries over a particular collection." Although the discussion of federated search is beyond the scope of this book, it is generally accepted that figuring out the person's goal is more important than allowing them to select a particular collection because people who are not skilled librarians usually have a hard time differentiating between various databases. If you're interested in the topic, pick up a copy of Peter Morville's essential book, *Search Patterns: Design for Discovery* (2010, O'Reilly).

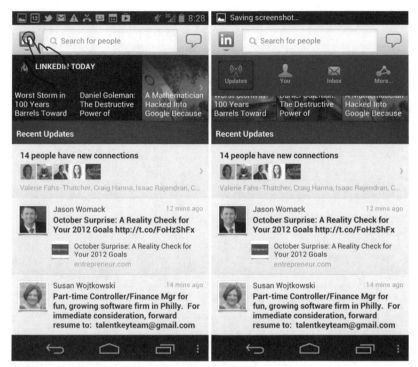

FIGURE 13.10: LinkedIn app uses the Popover Menu pattern for navigation.

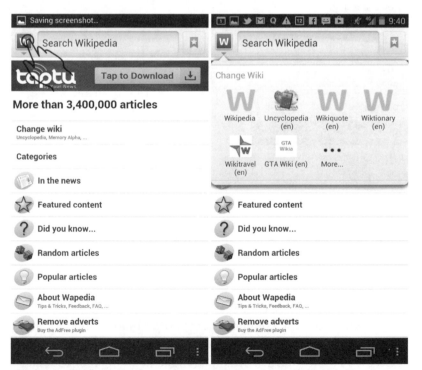

FIGURE 13.11: The Popover Menu pattern is used for selecting a search domain in the Wapedia app.

Finally, Figure 13.12 shows an example of a popover menu used to expand row-level functions in the Fandango app. Each row of the results includes a gray down arrow that opens a popover to reveal two additional functions, making a record total of three possible row-level alternatives for a customer to choose from.

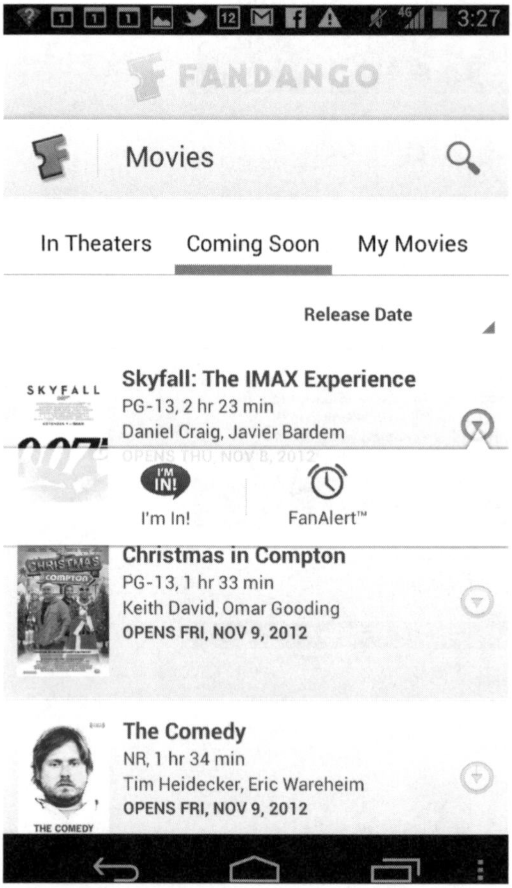

FIGURE 13.12: The Popover Menu pattern is used to expand row-level functions in the Fandango app.

When and Where to Use It

Any time you need to have more actions available to the customer than the confined mobile space allows, use the Popover Menu pattern.

Why Use It

Mobile apps have rapidly increased in complexity since their launch only a handful of years ago. Customers demand to have access to as many (if not more) functions on their mobile devices and tablets as are available on the desktop applications and desktop websites. This preference, with the fat-finger pointers humans are forced to employ, leads to a crisis of real estate. Showing more actions, filters, and navigation options becomes simply impossible in a confined mobile screen. Popover menus offer a flexible, extensible, and elegant solution to the problem by multiplying the real estate's capabilities through reusing space and opening a layer on top of the existing content.

Other Uses

There are simply too many popover menu uses to mention here. One important application is map-based navigation, where tapping a pin on the map brings up the popover menu. Another one worth mentioning is the Android navigation menu that comes standard with many older pre-4.0 Android apps. This menu had been the differentiating factor of the Android platform from the days of its inception, and it is the one popover menu whose evolution you definitely don't want to lose track of (see the later "Caution" section).

An alternative way to introduce row-level function expansions comes from Twitter. A tap-and-hold gesture on the individual row changes that row into a menu of actions (see Figure 13.13). Scrolling the page again dismisses the menu.

FIGURE 13.13: The Twitter app uses a tap-and-hold overlay menu variant to expand row-level functions.

This functionality is undoubtedly advanced, meaning that it has serious discoverability issues, and most people using the Twitter app have no idea that this function exists. It is also safe to say that a large number of those who do use the function have discovered it accidentally. This is largely because tap-and-hold menus are virtually unknown in Android; none of the native functions work this way, so the Android customers don't have a chance to learn the action. Nevertheless, in my opinion, this twist on the popover is by far the finest way to present row-level functions. It is elegant, minimalist, and authentically mobile, and it uses the confined real estate in the most effective manner. It would be a shame if the Google-Apple patent wars were to prevent this pattern from gaining universal acceptance.

Pet Shop Application

For the Pet Shop app, the most promising application of the Popover Menu pattern is the row-level action menu that enables actions such as Contact the Seller, Add to Favorites, and Share. Instead of using the Fandango app's model, however, the in-row menu variant introduced by Twitter is used—the tap-and-hold variant (see Figure 13.14).

In some apps, the action needed to call up the menu is changed from tap-and-hold to row-level, right-to-left swipe, which is also borrowed from the iOS multitouch toolkit; although they aren't discussed here.

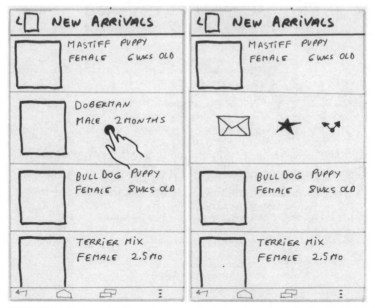

FIGURE 13.14: The tap-and-hold overlay menu variant expands row-level functions in the Pet Shop app.

Tablet Apps

Tablets do not suffer as badly as their smaller mobile counterparts from the lack of real estate used for navigation. Tablets are also more conducive to multitouch explorations, as observed in many field studies. This is mainly because of their "gaming DNA"; experienced tablet owners are more likely to expect funky and unusual multitouch behaviors from their devices and are more apt to look for these gestures during work and play.

There's another unspoken expectation that I have dubbed "Is this all there is?" Most owners expect that a tablet will have greater functionality than the mobile device, even if it is just shortcuts and action accelerators. Regardless of the device, however, the discoverability of the "hidden" navigation options, such as row-level functions, is always a challenge. See the next pattern, Watermark, for ideas on how you can communicate the action necessary to unlock the "hidden" multitouch functions to your customers in a virtually foolproof way.

! Caution

Make sure to consider all menus. In the earlier LinkedIn app example, the pop-over menu is used for navigation. Unfortunately, this makes the Android device bar menu almost obsolete because it now displays only a single function: Refresh under the Updates tab. Worse yet, under the You tab, the Android device bar menu is completely nonfunctional, and tapping it shows no options and has no effect on the app (see Figure 13.15).

This behavior is confusing and contrary to both the Android data model and customer expectations.

Also, don't mix selection with navigation. In the earlier Wapedia example, the popover menu is used for selecting the collection for searching. However, it has one additional use: navigation to the More section. Navigating to the More section retains the search box and resets the collection choice to Search Wikipedia, which is somewhat unexpected and confusing because it is at odds with the previous options, essentially mixing the navigation options with filtering in the same menu (see Figure 13.16).

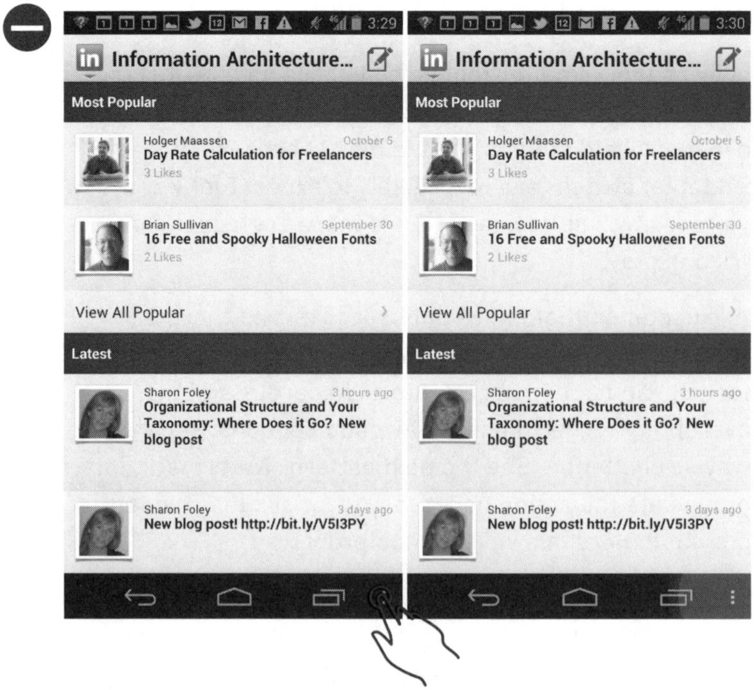

FIGURE 13.15: An antipattern occurs in the LinkedIn app where it strongly deprecates the Android device menu in favor of the popover menu.

FIGURE 13.16: This antipattern in the Wapedia app popover mixes filtering with navigation.

Do these issues signal the end of the world as you know it? No, of course not. However, there is a much better model in the native Android Google apps, as exemplified by the latest incarnation of Google Plus, which removes the navigation bar Android menu in favor of using two top menus in the action bar (see Figure 13.17). Much like in the LinkedIn example, the top-left action bar menu is used for global navigation, whereas the additional Android overflow menu in the top-right corner is used for section-level and global functions, such as Search and Settings. The more commonly used functions such as Refresh and Compose are given dedicated icons in the top action bar (see more about the Drawer element in the Google Plus app in Chapter 1, "Design for Android: A Case Study").

FIGURE 13.17: The Google Plus app divides and conquers navigation and functionality with two popover menus.

This is an elegant and novel model that is clearly and loudly signaled by Google as a departure from the traditional bottom navigation-bar menu that had been the Android mainstay feature from day one. When using the popover menu for your own navigation, consider which options need to be included in which menu, and be sure to test various options with your target customers.

Related Patterns

13.5 Pattern: Watermark

13.5 Pattern: Watermark

Watermark provides the essential hint that enables customers to discover "hidden" multitouch gestures and accelerometer actions that can be used for navigation.

How It Works

When the multitouch gestures and accelerometer motions exist on the system, the screen shows a semitransparent watermark (in some cases shown underneath the content). The Watermark pattern provides the hint of the gesture or motion that the customer needs to perform for a specific system action. It is shown briefly and then melts away. Usually the sequence is repeated several times at each start of the app, or until the customer performs the action the first time. After that, the watermark appears at regular intervals (biweekly for example) as a reminder if the system detects that the multitouch gesture/accelerometer motion has not been performed in a while.

Example

Games provide great examples of cutting-edge patterns. Watermark is no exception. For example, in the delightful politically incorrect game Major Mayhem, The Major shoots at ninjas in a jungle. At the start of the game, a free-form animated contextual watermark shows how to use your finger to perform two functions: the basic tap-to-shoot, and the more complex motion of tap-and-drag (see Figure 13.18), which drags poor ninjas from behind obstacles and defenses.

As the game progresses, more actions, such as jump, are demonstrated for the player until she learns all of the available multitouch gestures.

Chapter 5, "Welcome Experience," already touched on this subject of contextual tutorials. What makes the Watermark pattern different is that it is not an overlay that you need to dismiss; instead it's *a gentle invitation*. Watermark is completely integrated into the context of the current task, so the customer can simply proceed with the task without following the motion or action suggestion in the tutorial. The customer retains her freedom, and the watermark is simply there to remind the customer that the action can be performed. In the Major Mayhem example, the customer does not *have* to perform the tap-and-drag action of pulling the ninja out from behind the obstacle; the person is free to shoot the ninja instead. In this way, the watermark is unobtrusive enough not to interfere with the primary task (shooting those pesky ninjas).

When and Where to Use It

Any time you employ gestures other than the simple tap, it's a great idea to use the Watermark pattern to help your customer discover and remember the gestures and avoid any confusion.

FIGURE 13.18: Here the Watermark pattern is demonstrating in-game gestures in the Major Mayhem app.

Why Use It

Multitouch and accelerometer offer a fantastic array of both natural and novel gestures that can help your customers navigate complex applications on mobile devices, without adding bulky navigation options. Unfortunately, these gestures are largely undiscoverable without some sort of a hint. Watermark provides a playful hint, a gentle reminder about the available actions while still allowing the primary task to proceed. After you decide to use a Watermark, it is tremendously freeing and beneficial to the design of the app and the resulting customer experience because it instantly liberates your team to explore alternative, more compact and elegant, authentically mobile navigation layouts. Patterns like Watermark are also more likely to help create simple, uncluttered displays necessary for immersive experiences—the subject you explore in the next section.

Other Uses

Watermark is not just for games; see the "Pet Shop Application" section for some examples of how to use this pattern for "serious" e-commerce and social media applications.

Pet Shop Application

For the Pet Shop app, imagine that a simple filtering action can be accessed via a shake, as in "shaking up the search results," instead of using a dedicated Filter button. A simple way to show this would be with a Watermark pattern that shows the shaking action (see Figure 13.19).

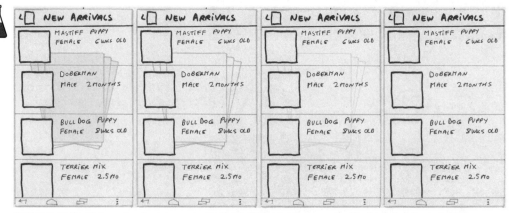

FIGURE 13.19: The Watermark pattern demonstrates the shaking action for accessing filtering and sorting in the Pet Shop app.

Note that the wireframes show the watermark melting away after a short while both to remove the potential annoyance and to draw attention gently to the invitation for the additional action. Remember, you don't need to actually show what will happen when the shaking occurs, only to reminder customers gently to try it out and allow them to discover the result on their own, and so naturally fall in love with your app.

Another novel way to perform the filtering action utilizing a multitouch gesture might be to draw a circle on the screen with a finger—perhaps reminiscent of the motion of mixing marbles (results) in a bag (results set). The wireframe in Figure 13.20 shows how this might look using the same set of results.

FIGURE 13.20: The Watermark pattern in this case demonstrates drawing a circle to access filtering and sorting in the Pet Shop app.

Notice that none of the watermarks in this section interfere with viewing the results. They merely hint at actions the customer might want to try out to access additional functionality.

Tablet Apps

See the "Tablet Apps" section in the "13.4 Pattern: Popover Menu."

! Caution

The customer will expect that if she does not perform the action suggested by the watermark, it might go away and then re-appear at some point as an additional reminder. However, when the customer performs the action, it's expected that she will not see the watermark again for quite some time (think weeks, not days) and that eventually it might go away altogether.

It is an art of how to remind without being annoying. The design field is littered with loathsome examples of annoying tutorials, such as Microsoft's Clippy, as shown in Figure 13.21 (image credit: Mark Stanger, "Why Service-now.com won't ever have an animated help assistant," December 29, 2009, http://www.servicenowguru.com/service-now-miscellaneous/servicenowcom-animated-assistant).

It looks like you're trying to get some work done. Would you like me to bug you instead?

- Annoy me till my eyes bleed

- Go away please, but come back right in the middle of my PowerPoint presentation this afternoon.

FIGURE 13.21: Microsoft's Clippy was possibly the most universally hated contextual tutorial agent of all time.

Although the Watermark pattern has the general advantage of being more unobtrusive than most, it is always better to err on the side of giving too little help rather than too much, particularly for animations, including appearing and disappearing, which tend to draw a lot of customer's attention. You might want to have a built-in dead-switch where the watermark goes away after being shown a total of four or five times, or remind people at most once a month. You may even want to consider having a settings (On/Off) switch, such as "Display reminders?" It is a safe bet that if the customer performs the action on her own within 10 minutes of using your app, she has learned the action by heart and does not need to be reminded again for at least 2 weeks. Regardless of the strategy you choose, make sure you test the Watermark interaction thoroughly with your target audience.

Another thing to remember is not to over-indulge; reserve multitouch gestures for unique, special actions the customer might take once in a while. For actions that require multitouch as the primary action, be sure to provide a simple push-button alternative. For example, a popular app called Urbanspoon uses the gesture of shaking the device to communicate to the system the action of searching. From the first glance this interaction was cool, and indeed, this unique feature did draw a great many people to download the app initially. However, people quickly discovered that shaking every single time you need to search is actually tedious, so designers were quick to add a Shake button (see Figure 13.22) that performs the same action. This shortcut turned out to be the magic sauce to success: Draw the customers in with the cool multitouch gesture but provide the back-door access to advanced users who are familiar with the system.

FIGURE 13.22: Urban Spoon provides back-door access to its shaking gesture.

In general, on mobile devices, consider having push-button, back-door access for your key multitouch gestures that are repeated frequently. Place the button in a location that is ergonomically convenient to access with a one-handed, right-hand grip. (See the device ergonomics section in Chapter 3, "Android Fragmentation" for details on where the convenient access "hot zone" is located on various touch devices.) It's a great idea to label the action with the name of the multitouch gesture, such as Shake, Swipe, and so on. This can also serve as a reminder of the multitouch action. After the novelty has worn off, the button is likely to be the primary way your customers access the action.

Related Patterns

5.5 Pattern: Tutorial

13.6 Pattern: Swiss-Army-Knife Navigation

If you ever played a game on your mobile device, you know that navigation generally recedes into the background in favor of lending the entire screen to the primary task of entertainment and challenge. This pattern shows you how to adopt this type of Swiss-Army-Knife Navigation to any application.

How It Works

This pattern is all about exploring the experimental trends that use navigation that recedes into the background so that the content can shine. Most of the pattern implementations accomplish this with the hidden navigation menu that opens by tapping a button in the corner of the device. In the various modifications, navigational menus come down as window shades, slide the content over, appear in the overlays, and use all manner of creative elements and transitions. In games, the menu button is frequently semitransparent, allowing for the entire screen real estate to be used for play.

Example

A favorite example of a successful immersive experience is the game Angry Birds. With more than a billion copies downloaded (according to Wikipedia—http://en.wikipedia.org/wiki/Angry_Birds), Angry Birds is a model for a successful immersive experience that uses the Swiss-Army-Knife Navigation pattern.

FIGURE 13.23: The Angry Birds app is an excellent model of an immersive experience with its semi-transparent button and window shade menu.

The entire area of the screen is devoted to the immersive task of squashing the pesky pigs. The only navigational control is the semi-transparent round pause button that opens the custom window shade menu that flies in from the left side of the screen. Note that the menu is highly customized to appear as part of the game, and when it flies in, the menu appears to come on top of the game's content, keeping the player immersed in the flow of the game and maintaining the context. To accomplish this, the menu does not cover the entire game area of the screen, and the main area of the screen is subtly darkened during the smooth transition. When creating an immersive experience, seemingly small elements like shading, icons, and transitions make the key difference.

Another window shade navigation experience is offered by the Wells Fargo Bank app (see Figure 13.24). It is remarkable because although the app is a hybrid app (that is, it uses HTML elements in a captured browser window), the window shade transition is smooth and believable. This app is a tribute to both the developers and designers as well as the capabilities of the HTML 5 apps in general.

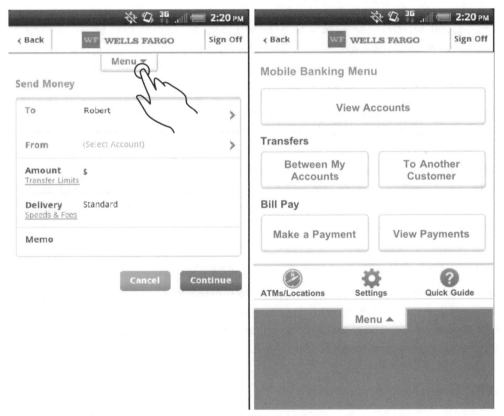

FIGURE 13.24: Here's the Swiss-Army-Knife Navigation in the Wells Fargo app.

Despite the masterful execution of the menu element, the quality of the stage set for having an immersive experience does not quite match that of the Angry Birds app because of the additional navigation elements present in the app. But maybe it does not have to; banking apps are different from games. Instead, the Swiss-Army-Knife Navigation menu button accomplishes its purpose of providing Wells Fargo customers with reliable, complete navigation access throughout the entire app.

The lesson here is that although the main purpose of the Swiss-Army-Knife Navigation pattern is to devote more of the screen real estate to crucial content, the pattern also offers significant fringe benefits of providing universal navigation throughout the app. Universal navigation makes it possible to completely dispense with the need for the List of Links pattern discussed previously in Chapter 6, "Home Screen." Instead, when freed from the burden of navigation, the homepage (and indeed, the entire app) can be made relevant to the customer through patterns such as Updates and Browse, showing local, relevant information and deals.

One of the apps that takes full advantage of the immersive navigation capabilities is Facebook, which recently redesigned the global navigation using the side menu teaser variation of the Swiss-Army-Knife Navigation pattern (see Figure 13.25). Tapping the top-left corner opens a sliding menu that comes in from the left side, leaving just a teaser of the main content on the right side of the screen.

Many designers and developers argue endlessly about the differences between the Angry Birds and Facebook implementation of this pattern, but you, the gentle reader, can dig deeper and recognize the common design DNA of both approaches. No matter if the menu has a slick transition or not, and whether the menu is scrollable or if it overlays the content as it does on Angry Birds or pushes it over as in Facebook, both have the same purpose in mind: giving content center stage while allowing universal navigation from every screen in the application. That's power.

When and Where to Use It

Use this pattern any time you need to devote more of your real estate to content, and promote the state of flow in your customers' activities. You should also use it when you want to show more of the content and less chart junk (navigation) or when your customers are savvy enough to figure out how to get to the navigation.

Why Use It

Swiss-Army-Knife Navigation promotes flow and immersive experiences. This is especially true on the small mobile screen, where the real estate devoted to navigation usable by fat fingers encroaches on and distracts from the real estate that can be devoted to usable content. Swiss-Army-Knife Navigation provides an elegant approach to removing navigation from view until it is needed and then devoting almost full-screen real estate to it when the customer decides to navigate somewhere else.

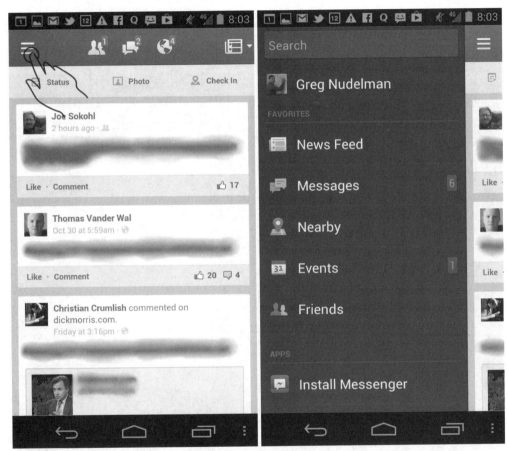

FIGURE 13.25: The Facebook app uses excellent Swiss-Army-Knife Navigation.

In the words of Marissa Mayer, "Google has the functionality of a really complicated Swiss Army knife, but the home page is our way of approaching it closed. It's simple, it's elegant, you can slip it in your pocket, but it's got the great doodad when you need it. A lot of our competitors are like a Swiss Army knife open—and that can be intimidating and occasionally harmful." ("The Beauty of Simplicity," *Fast Company*, November 2005). The Swiss-Army-Knife Navigation pattern is about translating this same sentiment to mobile navigation.

Other Uses

Swiss-Army-Knife Navigation doesn't need to always be implemented as a button that causes the menu to fly out. Other models have also been proven successful for a variety of specialized applications. For example, while viewing a photo in a native Photo Gallery app, a single tap in the middle of the image brings up a semi-transparent overlay with various functions, such as sharing, e-mailing, deleting, and so on (see Figure 13.26).

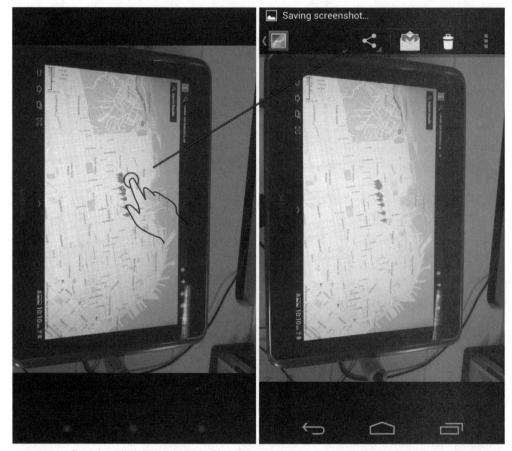

FIGURE 13.26: Swiss-Army-Knife Navigation in the Photo Gallery app uses a single tap in the middle of the screen instead of a button.

This same overlay never appears in the viewing mode of swiping through the image collection, so the navigation stays hidden, and the customer remains in the flow of viewing the images.

Of course, this is a specialized application of this design pattern that works only because the viewing mode uses multitouch gestures. It does not work for most apps that require tapping somewhere in the screen. But maybe the inverse approach might work. For example, you might consider retracting the navigation while in the flow mode and using some special multitouch gesture to bring it back. This is exactly how the native notification function works on the Android platform. Notifications are almost completely hidden at the top of the screen. Swiping down brings up the menu as shown in Figure 13.27. (Curiously, this exact multitouch motion is used by Windows RT to return to the home screen.)

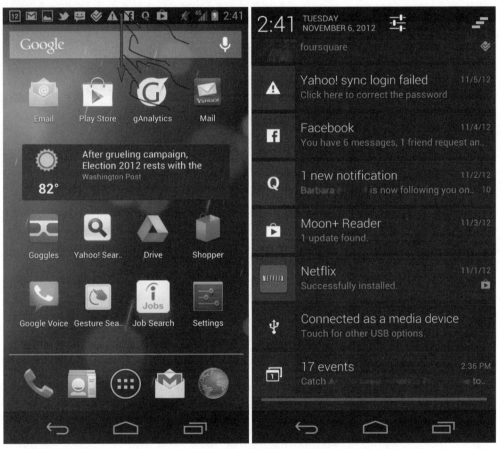

FIGURE 13.27: The Swiss-Army-Knife Navigation pattern is used to bring up the Android Notifications screen.

As the most popular top-to-bottom swipe is taken by the native notification control, you might want to consider using a side-to-side swipe (popular in

Windows Modern UI), diagonal swipe, or shake instead. Another alternative is to use a touch screen to draw a letter like "G" for "Go" or "M" for "Menu."

 Be sure to also consider and experiment with "made up" gestures that are easy to draw. A game called Infinity Blade uses touch screen gestures to cast spells quickly in the midst of battle. As author's ample experience with this awesome game shows, a simplified gesture similar to the Cyrillic letter "Gh" (that looks like the inverted "L") is one of the easiest and quickest to draw, as is letter "C" (half-circle) in any orientation, whereas drawing a "U" reliably and quickly is difficult (see Figure 13.28).

FIGURE 13.28: These novel multitouch gestures are from Epic's Infinity Blade game.

Finally, if you can't remove the navigation entirely, consider making it semitransparent instead, as exemplified by Google Maps and Google Earth, as shown in Figure 13.29.

Although both action bars are always displayed on screen, they are semitransparent, which enables customers to see the map contours under the bars. This is a good intermediate solution if you cannot go all the way to creating a Swiss-Army-Knife Navigation because you can still take advantage of some of the benefits offered by this pattern.

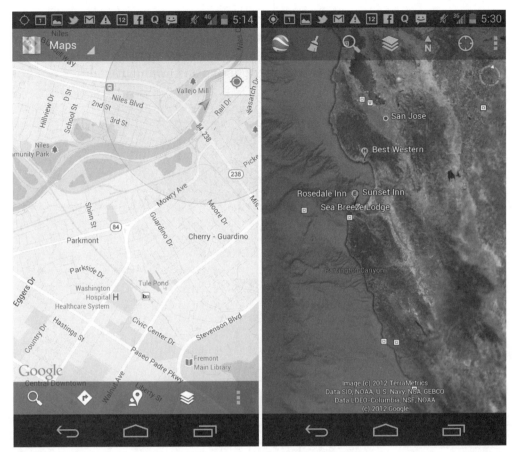

FIGURE 13.29: The Google Maps and Google Earth apps use semitransparent action bars.

Pet Shop Application

For the Pet Shop app implementation of the Swiss-Army-Knife Navigation pattern, see the wireframe in Figure 13.30 for a four-corners pattern variation. The idea is simple: Instead of using only a single semitransparent menu button as in the Angry Birds game, the app uses all the corners of the screen as partially transparent shortcuts to key areas of the app. For example, in the search results shown, the two top spots are Map and Shopping Cart, whereas the bottom-left corner is occupied by the Filter and the top-right corner by the menu.

The corner shortcuts are also combined with the semitransparent Filter Strip element (see the "8.3 Pattern: Filter Strip" section in Chapter 8, "Sorting and

Filtering"). Together, this design enables 100 percent of the screen real estate to be used to scroll through the cute pictures of puppies. At the same time the key areas and capabilities of the app are well exposed, all without the slightest possibility of accidentally hitting a button and triggering an unwanted action.

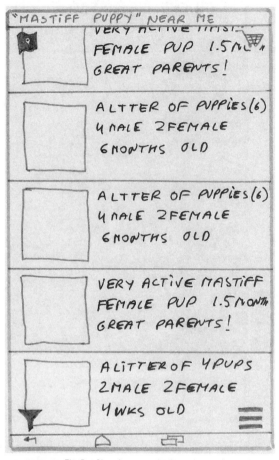

FIGURE 13.30: The Pet Shop App uses the four-corners Swiss-Army-Knife Navigation variant.

Why this choice of placement? Placing the menu in the bottom-right corner makes sense—this is where the Android navigation menu traditionally rested in older apps, so it makes sense there when you replace it with a custom window shade. Also it is easily accessible by the right thumb in a right-handed one-hand grip. The Filter button is on the bottom left because it is the most likely button to be used again and again in the context of a particular search. Ergonomically, Filter is in the easiest corner to access because the thumb naturally rests in that area in a one-handed, right-hand grip that you see commonly in mobile devices. The top-corner buttons are there mainly to raise awareness of the features that may not be immediately top-of-mind, such as Map, and to house the ubiquitous Shopping Cart button. Although this arrangement makes sense, many other arrangements with various functions are possible. Experiment with your own designs to determine what works best. The key is to try to devote as much space as possible to content while at the same time giving access to key functionality.

Tablet Apps

For larger 10-inch tablets, this space-saving pattern is not needed because there is much more space available for navigation, toolbars, and the like. However, for 7-inch tablets it should be definitely considered. High-end visual adaptations of the four-corners design variations are also fun to design and use. For example, you can imagine a slick text editor devoid of extraneous controls, menu bars, and buttons—a blank canvas on which to write the next great American novel. The bottom-right corner of the canvas could be subtly raised (as though it is a physical piece of paper). This corner, when tapped, can be used to call up the menu which is revealed behind the screen by rolling up the canvas in a smooth transition.

! Caution

My biggest caution here is not to wait to use this pattern before your competitors do! If you are worried about discoverability of the hidden Menu button, don't be. All of the several billion people who downloaded *Angry Birds* and other games that use the same pattern learned to use it successfully. In a study sponsored by a telecom giant, more than 25 people revealed that when presented with only a single action button, people will click it and discover the meaning behind it experimentally, even without full understanding of what the button does. With two or more buttons, things do get trickier, so make sure to use good icons and test the interface early and thoroughly with your target audience.

Remember, you don't need to use all four corners. For example, the Facebook app (see Figure 13.31) uses the two top corners plus a native Android navigation bar menu on the right-bottom side of the screen (which makes three corners total). The design is slightly different from the one presented in the Pet Shop app, but the general approach presents an excellent paradigm for your own experimentation.

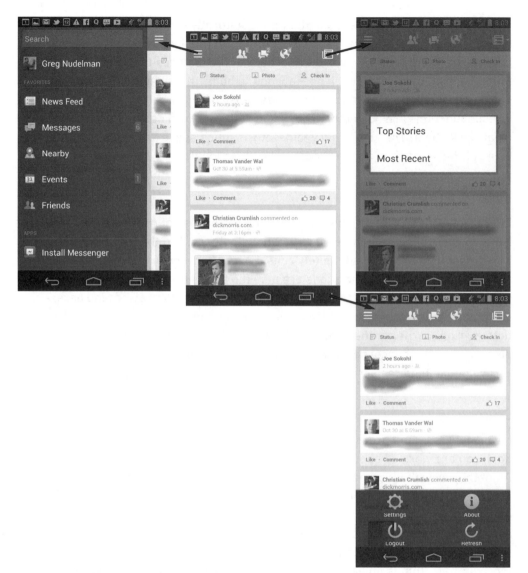

FIGURE 13.31: The Facebook app uses a Three-Corner Swiss-Army-Knife Navigation pattern.

You might prefer to have icons that are a bit more differentiated than those that Facebook uses. This is essentially a problem of figuring out which action is under which corner menu. When in doubt, try using one main menu and abstracting additional key functions only after the menu contents have been largely worked out.

Related Patterns

8.3 Pattern: Filter Strip

13.7 Pattern: Integration: The Final Frontier

This final mobile pattern departs somewhat from the single-app theme, here-tofore used in this book in favor of showing how navigation looks when control is passed between the apps. It also discusses exciting widget and OS-level opportunities for aggregating data from various apps into highly usable mobile dashboards.

How It Works

When the needs of your customers exceed the capabilities of your app, you need to navigate to a different app that takes over the task for the customer. Besides the purely procedural integration, opportunities exist to offer integration of data feeds from various apps and display them on custom dashboards, widgets, and panoramas that serve as jumping off points for engaging with these apps.

Example

One example of a seemingly simple procedural integration is the FourSquare app. The app offers a "captured" version of the Google maps that is licensed specifically for displaying the list of places of interest in the map format. However, to obtain directions, the customers must navigate to the separate native Google Maps app (see Figure 13.32).

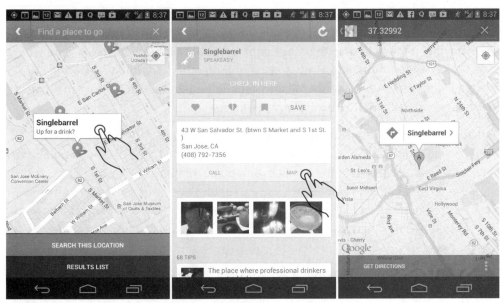

FIGURE 13.32: The Foursquare app uses a process-driven integration with Google Maps.

Foursquare is fairly well integrated. It passes to the Maps what appears to be a decimal number, which actually is a destination code. The Google Maps app correctly obtains the name and the address of the place of interest, presumably directly from Google's database. (Other integrations are not so lucky—see the later "Caution" section.)

When and Where to Use It

Use this pattern any time your app needs to provide services that are already well developed in a different app with which the customer is familiar.

Why Use It

Life is simpler when you don't reinvent the wheel. Also, there are excellent opportunities to improve the mobile experience and add value in pulling the data from various apps and aggregating it in an easy-to-scan format (see the next section).

Other Uses

One of the biggest opportunities in mobile is to offer skillful aggregation of data feeds from various apps. The main reason you do not yet derive full value from mobile apps is that communication is tightly subdivided into technological silos—e-mail, instant messaging, LinkedIn, Twitter, Facebook, and so on—and information is not semantically labeled and sorted into buckets people can actually use. Instead, people are forced to continuously check various communication channels for fear of losing an important communication and to get the constant dopamine hit of staying in touch and up to date. Keeping up with all methods of communication takes a lot of time and almost superhuman effort. Companies wanting to make advances in mobile communication should make a serious effort to separate the message from the technology medium.

One approach would be to integrate and prioritize various information feeds within a single unified inbox. Rather than switching apps to stay in touch on all the networks, you can get all of the feeds in one place. Instead of tracking down the people you need to get in touch with, you can select everyone you need and send a single communication to the right people in one shot, from one central location. Even better, people you want to reach could receive your communication in the way they prefer.

The way Android allows this kind of "integration" currently is purely accidental and unmanaged, via various widgets placed on the home screen (refer to Figure 13.33).

FIGURE 13.33: It's an antipattern that Foursquare and Twitter widgets can't be integrated on a single homepage screen.

Although this works well in theory, in practice, unfortunately, such widgets fail to meet the rising demands of people coping with the social networking feed information coming out of the internet firehouse. Even just looking at the FourSquare and Twitter widgets, there is a great deal of whitespace, little or no flexibility of what displays in the widget, and the sheer impossibility of putting both widgets on the same page because of their size. If the widgets were more integrated, you could fit Foursquare, Twitter, Facebook, and more on the same page. Individual sections of this mega-widget can function either as links or as sections that open to reveal more social feed information.

One intriguing opportunity to accomplish this on mobile devices and tablets actually came from the Windows Phone Panorama control, like the one shown in Figure 13.34 for Foursquare (image credit: Liz Ngo Global ISV, "Updated Announcement: Official foursquare 2.X App for Windows Phone Unveiled and Now Available for Download," MSDN.com website, June 28, 2011, http://blogs.msdn.com/b/msftisvs/archive/2011/06/28/official-foursquare-2-0-app-available-for-windows-phone.aspx).

FIGURE 13.34: The Foursquare app on Windows Phone uses this Panorama Control.

Every day new and different ways to communicate pop up all over the mobile land-scape. Facebook enables more than 10 different ways to send and receive mes-sages (poke, post on the wall, comment on photo, and so on.

Aggregating these various messages into a dashboard panorama makes sense. Unfortunately, it also comes up short of mobile technology's full potential. Instead of aggregating the information from various apps into a *single* panorama, vari-ous apps have created their own, individual dashboard panoramas, thereby again locking down the information silos.

This creates an excellent opportunity for agile and hungry Android product teams, who never shied from borrowing the most promising ideas from all kinds of sources. Imagine a teenager's (and small business owner's) dream: an open cus-tomizable Android Panorama dashboard for various network feeds, pokes, wall messages, check-ins, and alerts. Or imagine a CRM-in-Your-Pocket: a powerful individual dashboard for each of your contacts that includes the name and picture; Twitter, Facebook, and Tumblr updates; Foursquare check-ins; the last conversa-tion you had with them; and any IMs and voicemails you exchanged translated via voice-to-text. Imagine just how connected, informed, and engaged your customers would feel in the lives of their contacts! This is entirely possible to have today, yet no one has made an effort to use the Integration pattern in this fashion.

The collective mobile experience doesn't need to be about the technology silos, but rather about the goal of connecting, and the system should take on the task of aggregating and prioritizing various communication channels to present just the right amount of information. The system should prioritize messages and expressly alert people only to the most important and urgent messages. Android is actually the best platform to pull off these integration touch points because of the open standards and the spirit of experimentation it embodies.

Pet Shop App

What kinds of apps can the Pet Shop app be integrated with? Certainly maps for finding and displaying pets near you and getting directions for pick up. It could also have integration for news, events, and happenings in the customer's area, specifically for the pet category of interest (such as Guard Dogs). In this last mobile pattern in the book, I would like to invite you to brainstorm and sketch the possibilities!

Tablet Apps

Tablets are the best media consumption devices that are practically made for integration, period. Widgets and apps taking advantage of various smart integrations and aggregate dashboards are going to be the next hottest growth area for tablet app developers.

! Caution

Interapp integration is wrought with peril. Even "simple" map integration in major industry players such as Yelp and Kayak is far from perfect.

Figure 13.35 shows the data Yelp passes to the map is a combination of latitude and longitude (which is standard in Google maps) instead of the street address. This creates problems, such as not actually seeing the address or the name of the business in the Maps app. The Google Maps app also does not "remember" the business address in its drop-down of recent destinations, making it difficult, if not impossible, to multitask or try out different locations and then go back to the previous address, which are all standard tasks that anyone making a reservation would expect to do. Anyone watching the tap-flows that involve the maps component would notice that the same destinations are accessed several times in a row by the same person, indicating a subpar integration experience.

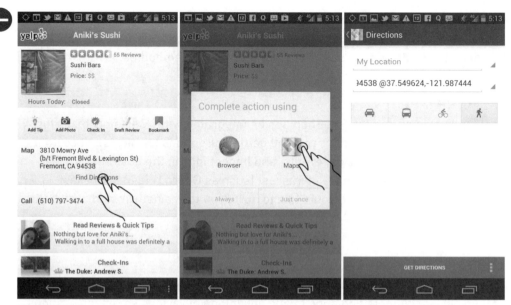

FIGURE 13.35: The coordinates-driven process-driven integration of Yelp and Google Maps has serious drawbacks.

You can achieve a deeper level of integration sometimes by "capturing" (or more correctly licensing) another app. One example of this is the Kayak app (see Figure 13.36), where the Bing maps app is used to display the location of the hotel on the map. Unfortunately, this integration looks better on paper than it does in reality. Just try to get directions to the hotel—you can't!

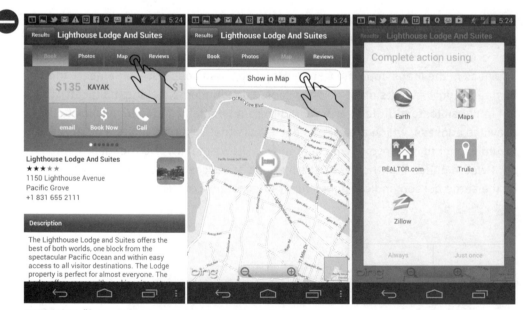

FIGURE 13.36: "Captured" Bing maps do not offer directions in the Kayak app.

Instead, the Bing-Kayak hybrid offers a funky-styled button labeled Show in Maps, which opens a puzzling array of apps (Google Earth, Zillow, Trulia, and Realtor.com) most of which use maps peripherally. Actually trying out those options yields errors and zero results, and for good reason: It appears upon careful examination that the Kayak-Bing hybrid passes "Lighthouse+Lodge+And+Suites," which you can see in Figure 13.37 comes up empty at (almost) every integration point (in the case of Google Earth actually failing without indicating that no results were found).

FIGURE 13.37: On Kayak, all integration points except Maps fail to find the target or provide directions.

Google Maps is the only one that actually yields something useful—the name and address located in the city of Pacific Grove—that can be used to obtain the directions originally desired by the customer. But why offer these other integration choices that do nothing but confuse customers?

Related Patterns

6.2 Pattern: Dashboard

9.1 Antipattern: Ignoring Visibility of System Status

CHAPTER 14

Tablet Patterns

Tablets are a huge area of mobile technology—these devices deserve their own separate book. The patterns included in this chapter are specific for small and large tablets with more screen area than a smartphone. Designing for tablets is one area in which a team willing to experiment can rapidly overtake the competition in creating a superior experience that is most closely suited to the unique tablet shape and device capabilities. Before tackling this chapter, review Chapter 3, "Android Fragmentation," for a refresher on the hot zones and most likely grip positions offered by various devices.

14.1 Pattern: Fragments

To address device fragmentation, Android raced ahead of the competition by offering the somewhat ironically named Fragments UI framework. Fragments forms the primary "pure Android" pattern of User Interface (UI) design.

How It Works

The content on the screen is broken out into tiles called *fragments*. Each fragment is positioned based on the device size and orientation to best occupy the available space. Scrolling of the individual tiles is limited so that essential elements, such as action buttons, do not scroll offscreen.

Example

My favorite example of the Fragments UI is the Google Play app. It is an example of a truly of a "responsive" native app that works surprisingly well on a variety of devices, as shown in Figures 14.1 and 14.2.

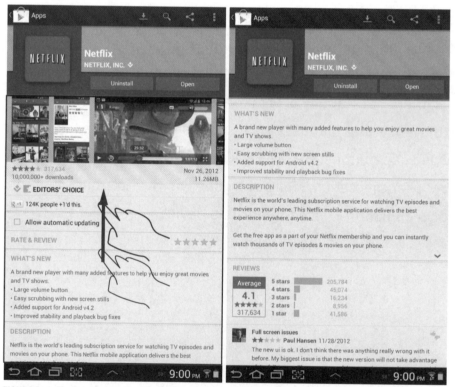

FIGURE 14.1: The Fragments pattern in the Google Play app forms a single column on a 7-inch tablet in the vertical orientation

In the horizontal orientation on the 7-inch tablet pictured in Figure 14.2, the elements of the Google Play screen flow into two columns as needed to occupy space most efficiently. When the tablet is in the vertical orientation, the elements stack instead into a single-column design as shown in Figure 14.1. Even though content within each of the elements scrolls independently, action buttons remain visible at all times. In each case the elements are arranged for optimal layout.

FIGURE 14.2: The Fragments pattern in the Google Play app forms two columns on a 7-inch table in a horizontal orientation.

When and Where to Use It

Translating existing app designs to small and large tablets is best accomplished using the Fragments framework. It offers plenty of capabilities and is easy to use.

Why Use It

The Fragments framework pattern offers the first serious stab at a one-size-fits-most approach. The Fragments pattern provides a decent experience on most devices while eliminating the need to create custom apps for various devices such as small and large tablets.

Other Applications

Much of what is written about responsive web design also applies to Fragments. However, the Fragments framework has been specifically optimized for responsive design, whereas much of the web design is done through CSS and JavaScript hacks. Read more about the Fragments framework at http://developer.android .com/training/basics/fragments/index.html.

! Caution

When laying out the fragment content for tiles that house the action buttons, limit the minimum size of the fragment to ensure that all the action buttons are shown to the customer, at least partially. This is important because a fragment with action buttons is difficult to scroll, and most people will not attempt to do so. Thus the owners of these devices will not discover the additional functions hidden below the fold of the fragment.

Be aware that no matter how great the intention, responsive design will not, in many cases, perform as well as a dedicated app designed specifically for a certain size of the device. If you are serious about providing a good tablet experience for your customers, you must take the time to optimize for certain popular sizes and test your Fragments UI on two or three popular device sizes in both orientations.

Related Patterns

14.2 Pattern: Compound View

14.2 Pattern: Compound View

Whenever the tablet is in the horizontal orientation, there is an opportunity to present the list with the detail element. This format is called the Compound View (also known as Two-Panel Selector, http://designinginterfaces.com/patterns/two-panel-selector/).

How It Works

Typically, on the smaller mobile devices, the list view is presented first and then the customer can drill down into the detail view. To get back to the list view, the customer has to tap the Back button. In contrast, tablets, especially in the horizontal orientation, are large enough to display both the list and the detail, making it easier and more efficient to drill into details without having to pogostick between the two views.

Example

A good example is the Settings app on the tablet device, as shown in Figure 14.3. The left side of the app's screen shows the list of settings, and the right side shows the details.

The Settings app makes an honest attempt to show the same information on both the vertical and horizontal orientations. It does this by performing a stretch/compress action, adjusting the column width of the left pane to achieve a balanced layout in both orientations.

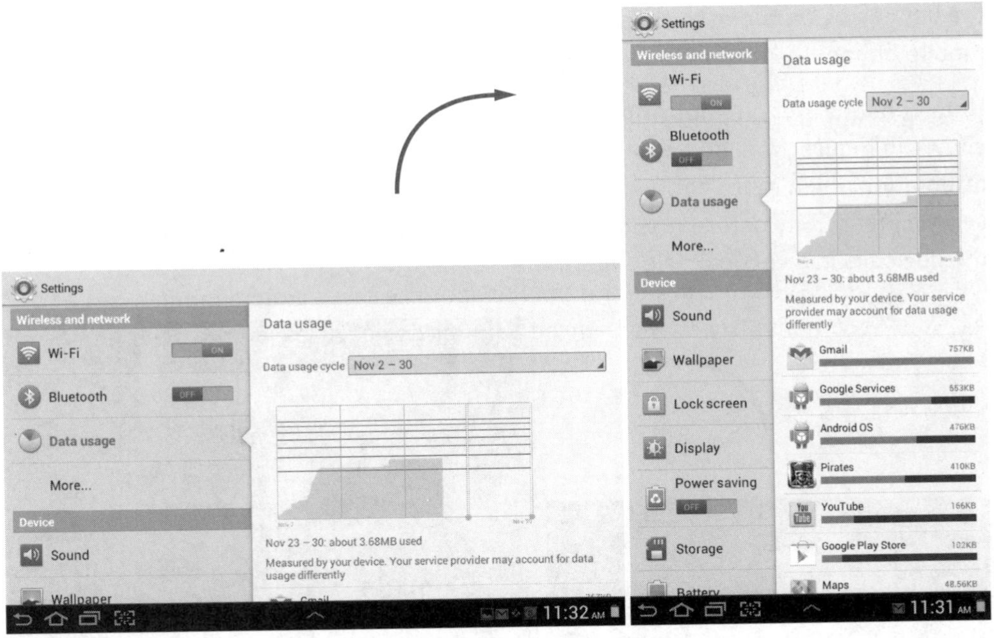

FIGURE 14.3: The Settings app shows an excellent Compound View implementation.

When and Where to Use It

Any time you have a list and detail view Information Architecture (IA) consider using the Compound View pattern.

Why Use It

Compound View is effective in giving the overview, while also allowing efficient random access navigation (that is, the person can navigate anywhere in the list view and see the detail without losing the context of the whole).

Other Applications

In some cases, the list column actually represents app navigation. In those cases, the Compound View pattern collapses into the useful and effective Side Navigation pattern covered in the next section of this chapter.

! Caution

Depending on the size of the device and the two columns of the compound view, some apps, such as the Gmail app, use a slightly different approach. In a horizontal orientation, they show the same two columns—list and detail—whereas in the vertical orientation they show only the detail column. In the vertical orientation, the list can be obtained by using the up navigation, as you would do while using a mobile phone.

Unfortunately, not all apps implement this correctly. Some, like the Ustream app shown in Figure 14.4, do not allow the customer access to the list view in a vertical orientation. Even more confusing, the column labeled Categories shows up on the screen just briefly before it slides out of view, without any way to get it back while the device is in a vertical orientation.

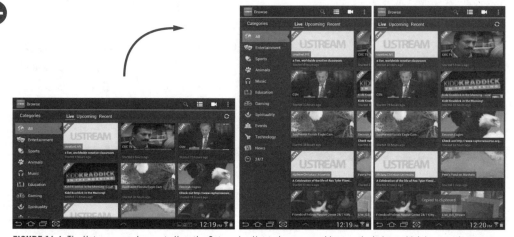

FIGURE 14.4: The Ustream app does not allow the Categories list to be accessed in a vertical view, which is an antipattern.

Fortunately, the customer can access the Categories list again by rotating the device back into the horizontal view. Despite that, this app design is an antipattern. To quote the Android guidelines, "Screens should strive to have the same functionality regardless of orientation." The keyword is *strive*—which means not all apps will have the same functionality in every orientation, and that is OK. However, when the constraints do force the list and detail screens to split, the list should always be readily accessible, and Ustream breaks that basic guideline, creating a very confusing experience.

Related Patterns

14.3 Experimental Pattern: Side Navigation

14.3 Experimental Pattern: Side Navigation

Side Navigation places the key functionality in the up/down orientation along the left and right sides of the device, where it is easily and ergonomically accessible without letting go of the device, which is particularly important for large tablets.

How It Works

Before deciding to follow the standard Android navigation guidelines, consider the ergonomics carefully, particularly for large tablets. As you recall from Chapter 3, for large tablets the sides of the device are more easily accessible than the middle of the top action bar. For apps that need to present a lot of navigation and functionality, running the functions up and down along the sides of the device in a compact menu makes a lot of sense.

Example

One example of a modified Side Navigation pattern is the Plume app. Plume is a popular, rich Twitter client for Android tablets. It displays the essential functions such as Home, Search, Favorites, and more in a Drawer element along the left side of the device (see Figure 14.5). Although it's not a bad implementation of the concept, you get the feeling that the design is not entirely complete, authentic, and minimalist. The Drawer menu is partially hidden most of the time, so the app must constantly remind the customer by moving the main content pane slightly, which jiggles the content periodically. There is also a great deal of other navigation along the top of the action bar and in the right side overflow menu, which makes the entire interface feel busy and incomplete, especially for a tablet that typically has plenty of navigational space.

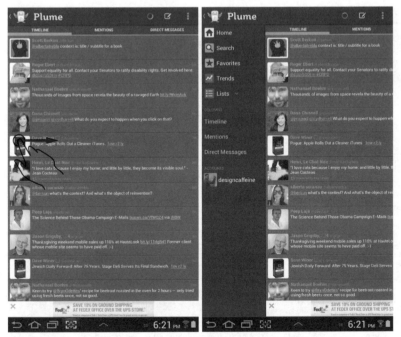

FIGURE 14.5: The Plume app includes a decent implementation of the Side Navigation as a Drawer pattern.

A cleaner implementation of the same functionality, which creates a better cus-tomer experience, is in the Twitter app on the Apple iPad. On the Android plat-form, this is, of course, an experiential implementation that nevertheless would work exceedingly well (see Figure 14.6).

FIGURE 14.6: An excellent implementation of the Side Navigation pattern is available on the Twitter iPad app.

The Twitter app was one of the first and most elegant and successful implementations of the Side Navigation pattern. The key feature of the left menu is that it *never disappears offscreen.* Depending on the orientation of the device and the position of the various panes of the application, the menu opens as a long list with icons and text, or collapses to just the thin, vertical row of icons by a simple, horizontal swipe. In contrast with the Plume app, there are no general icons and controls elsewhere in the app—*all* the functionality in the iPad Twitter app is located in the left menu.

When and Where to Use It

Any time you have a large tablet device, consider the Twitter iPad app style icons menu that remains on screen and does not completely disappear like the standard Android drawer. Depending on the goals and functionality of your app, consider this pattern for mid-size 7-inch tablets as well.

Why Use It

There are three reasons to use this pattern instead of the standard drawer:

1. This is a wonderful design because it enables the customers to learn the various functions quickly by looking at both the icons and the text. However, as soon as the customers learn the functions, they can collapse the menu to only the icons. If customers need a refresher, they can easily open the menu to show the text and be reminded of what the icons mean.

2. Placing all the app's functions in the single vertical icons bar running on the left or right side of the device lightens the cognitive load on the rest of the app.

3. All the functionality is located in the left vertical menu within easy reach of the thumb of the left hand, which means the customer can easily access any functions without having to let go of the device. This interaction works well in both device orientations.

Other Applications

Consider using the left nav icons in the vertical orientation if there is not enough space for the icons and text together. For instance, referring to Figure 14.4—how much more effective would the Plume app be if, after the initial load of the full two-column screen, the text would collapse, and only the category icons remained visible in the vertical orientation, while the full icons and text were available with a simple swipe? Figure 14.7 shows a hi-def wireframe of how this might look.

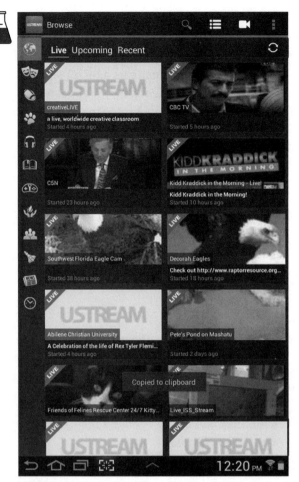

FIGURE 14.7: This proposed wireframe shows an alternative design of the Plume categories with icons that can slide out to a full menu.

In the horizontal orientation, the list can remain as-is or also collapse the same way, as needed, depending on the size of the device and the need to show more content in the right pane.

If you don't want to implement the icons in the left nav, at least consider using a simple bevel to show that the drawer is available and can be accessed by swiping, as in the AutoTrader case study in Chapter 1, "Design for Android: A Case Study."

! Caution

Caution here is of a political nature: Although it works extremely well, side navigation is not an accepted Android pattern—yet.

Related Patterns

13.6 Pattern: Swiss-Army-Knife Navigation.

14.4 Pattern: Content as Navigation/ Multitouch Gestures

Whenever possible, on tablets, use the content as a sweeping navigational element, avoiding tiny buttons that need to be tapped carefully.

How It Works

Every element of content is also a navigational element that can be accessed through various multitouch gestures.

Example

One of the smoothest examples of using content as navigation is Flipboard. (See Figure 14.8.) The app is not only beautifully designed, but it is highly functional as well, which leads people to truly fall in love with it.

FIGURE 14.8: The Flipboard app provides an excellent implementation of the Content as Navigation pattern.

The customer can access all the content simply by tapping on the large tile to see the rest of the story, swiping up to see more stories, and swiping right to left to go back. Even though there are no navigational buttons, few people have any trouble using the app because of its intuitive resemblance to a magazine.

A slightly different example comes from a creative app called News 360, which is shown in Figure 14.9. Each story is presented as a unique "cube" that can be rotated with an up/down swipe. A down swipe implies a "drill down" and reveals more information about the story. The up swipe reveals sharing functionality and additional navigation. The implementation adds a playful element through clever use of transitions and unique mapping of gestures to system actions that begs to be discovered and appreciated.

FIGURE 14.9: Playful content as navigation is implemented in the News 360 app.

However, simple directional swiping is only the beginning of the multitouch and gesture capabilities available on tablet devices. The Google Earth app (see Figure 14.10) enables people to interact directly with the real-world view they are looking at, creating an almost surreal virtual reality. Creative two-finger gestures such as Rotate and Pan are used effectively alongside the familiar stretch-and-pinch zoom function.

No one using the tablet version of the app fails to fall in love with the capabilities of the app or believe within minutes that this is "the only way to fly!"

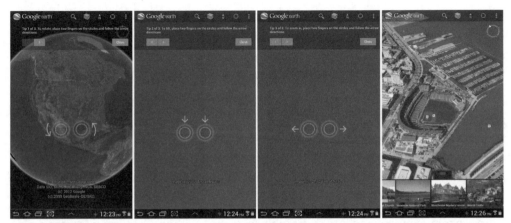

FIGURE 14.10: Complex multitouch gestures are used to manipulate the real-world content directly in the Google Earth app.

When and Where to Use It

Use direct-access navigation whenever possible for all tablet applications.

Why Use It

Tablets, especially large ones, lend themselves well to sweeping multitouch gestures. This has to do with the size and form factor of the device and also with the awesome and playful touch interactivity. Multitouch is a part of the two-handed committed use of the tablet and its touch DNA. As mobile User Experience expert Josh Clark says in his brilliant sidebar in my first book, *Designing Search* (2011, Wiley), "buttons are a hack." I couldn't agree more. Of course, buttons are still good for some commands that don't map to the basic set of well-known gestures, but overall, buttons on tablets are greatly overused and you should consider gesture and multitouch alternatives whenever possible.

Other Applications

As Richard Saul Wurman famously quipped at the 2010 IA Summit Keynote: "Everything we have done with computers and websites is still really primitive. We are in a really rapid changing time during which you can't invest in the finality and finish of anything. As a result most things are really bad. The web is not going to end up as a collection of pages. It needs to be a fluid movie and provide a journey. You need to be able to fly through the information of your choosing. You need to be able to direct where you want to go. The metaphor of the book for the web does not support that."

Substitute "information" for the word "web" and you have the model for how to view the future of the customer's relationship with your data. We are only just scratching the potential of direct manipulation of content through multitouch. The applications of this pattern are truly limitless.

Here's a simple, yet concrete, example of one way this could possibly work: Using a picture of a riveted boot, a customer should find similar boots by circling the features she likes (rivets), enlarging the features she wants to make bigger (as in a wider toe, for example), and literally crossing out the features she does not want (the heel). Consequently, the results the app suggests would be a bunch of riveted boots with a wide toe and no heel.

Other recent explorations into this highly immersive "flight through information" include rich tablet magazines such as *Opening Ceremony* magazine, and multi-media stories that include touch, such as *The Fantastic Flying Books of Mr. Morris Lessmore*. For these single-use touch gestures, remember to keep the touch portion of the story light and engaging, and the multitouch motions simple and intuitive so that people can recognize them using simple onscreen cues without having to engage in a long tutorial.

! Caution

Some gestures are not as discoverable as others. To avoid "mystery meat" navigation and undiscoverable functionality, use watermark tutorials, overlays, and animations to show interactivity, as Google Earth does so effectively (refer to Figure 14.10 and see the "Related Patterns" section).

Keep in mind that tablets do not lend themselves well to some accelerometer motions, even though the motions are "allowed." For example, gentle tilting is fine, but shaking and flipping is quite awkward, especially for larger, heavier devices.

Related Patterns

13.5 Pattern: Watermark

5.5 Pattern: Tutorial

14.5 Pattern: 2-D More Like This

2-D More Like This is a simple but powerful content browsing design pattern. It's particularly effective for larger touch devices such as tablets, where it can be used to transform a variety of search tasks into a pleasurable, visual browsing experience.

How It Works

Search results are placed in a gallery format across several rows, with each row representing a particular subdivision of the result set. Rows can be created from any division that makes sense, such as subcategory, brand, date, or price ranges. Each row is equipped with a carousel control (see the "13.3 Pattern: Carousel" section): In addition to a few thumbnails that can be displayed to the customers, each row also extends two or three screen widths to the right, which enables the customer to view additional elements in each row by swiping right to left. The customer can also scroll the page up to see more rows. The result is a two-dimensional scrollable matrix of thumbnails that are organized by topic.

Each row is also equipped with a More Like This link that's somewhere in the row so that no scrolling is required to select it. On touch screens, you can typically accomplish this by making the row title act as a link. For best results, the More Like This link could also be placed as the last spot in the scrollable carousel. That way, if the customer does not find what he is looking for in the carousel, he can tap the link to see more search results that match both his query and the topic of that particular row.

Example

One of the first apps to successfully use the pattern is Netflix, as shown in Figure 14.11.

FIGURE 14.11: The Netflix app includes an excellent 2-D More Like This pattern implementation.

This pattern works equally well in the vertical and horizontal orientations—one displays more rows; the other displays more items in each row. The customer can rotate the tablet to fit his needs for the specific task.

When and Where to Use It

You can use the 2-D More Like This pattern to support a variety of browsing tasks for content that is easily categorized. This pattern works best for result sets that are primarily visual, but it can include additional captions or even text snippets.

Why Use It

Tablets lend themselves to "contemplative consumption" of information accessed using large, sweeping gestures, appropriate to the larger size of the device, with minimal typing. There are few search design patterns better suited to taking advantage of this than 2-D More Like This.

Well-designed More Like This pages take full advantage of the full available surface area of the device, work well in both landscape and portrait, and provide excellent ergonomics. Overall, this pattern contributes to the feeling of flow: immersive, elegant, strain-free flight through visual information.

Other Applications

The 2-D More Like This pattern combines easily with other multitouch UI elements, such as slide-out menus. For example, in the elegant implementation of the 2-D More Like This pattern in the Pulse app that's shown in Figure 14.12, sliding the screen further left reveals a menu drawer.

! Caution

You should strive to avoid the following common pitfalls of the 2-D More Like This pattern:

1. **Don't get stuck with the same subdivision for each row on the page.** Although this is a common application of this pattern, it is by no means the only one available. 2-D More Like This is not just for displaying the subcategories of a single parent category. Instead, use subdivisions that make sense for your audience. One row can be subdivided by category, the second by brand, the third one by price range, and so on. People find this approach useful and practical, and do not get lost in trying to figure out the IA of individual rows. Instead, they dive straight in and begin using the information by exploring the rows that make the most sense to them.

FIGURE 14.12: This elegant 2-D More Like This pattern implementation with a menu drawer is from the Pulse app.

2. **Don't forget the teasers.** If you do use carousels in each row (which is highly recommended on touch devices) remember that the extra content is not entirely intuitive or discoverable. The best way to showcase that more informa-tion can be obtained by scrolling is by showing a *teaser*—a partial view of the

next item. You can use teasers on the right to show that there are more items in each row, and you can use teasers on the bottom of the page to show the customer that she can access more rows by scrolling down. Don't forget that teasers need to work well in portrait and landscape orientations. The best way to ensure that is to increase the number of rows or number of items across the row, respectively, based on the particular device orientation.

3. **Use real items.** Although you can use this pattern with all kinds of visuals, the best implementations of 2-D More Like This use images of real items (thumbnails) in each row, not icons or drawings.

4. **Use the smooth scroll inertia.** Some implementations of the 2-D More Like This pattern use the alternative "one page at a time" scroll behavior for scrolling horizontally across a row. This isn't recommended. To maintain the feeling of flow, each row's carousel needs to have the same smooth scrolling inertia as the rest of the page, ideally adjusting the initial scroll speed of the row in response to the speed of the horizontal swipe gesture.

Related Patterns

13.3 Pattern: Carousel

14.6 Experimental Pattern: C-Swipe

If the recent Windows 8 developments are any guide, tablets are going to get larger over time. Already the lineup includes 12-, 15- and 21-inch touch screens, and touch-friendly applications are becoming more complex and full-featured. It's only a matter of time before Android will be forced to catch up, yet scaling the current action bar scheme may not be realistic or ergonomically desirable. C-Swipe is a futuristic, experimental pattern that forms the basis for an alternative navigation scheme. It can be used to bring up a contextual menu anywhere on the touch screen using a natural semi-circular arc described by the human thumb along the surface of a flat touch screen. This gesture is roughly in a shape of a letter C, hence the name of this pattern.

How It Works

Imagine this: The entire surface of the tablet is devoted to content. To use functionality or navigation, the customer swipes the screen with her right thumb in a natural semicircle gesture near the edge of the device. She can make the gesture while holding the device comfortably and securely with her wrist and the remaining four fingers of her right hand and the entire five fingers and a palm of her left hand.

The swipe of the thumb causes a semicircular contextual menu to display. The most commonly used function is on the top, near the final position of the thumb following the completion of the C-Swipe gesture. Icons and associated text of the menu are positioned in such a way as not to be blocked by the thumb. The person taps the function she wants, and the menu once again disappears as the action is performed.

Example

As mentioned earlier, Flipboard is an elegant app that uses multitouch effectively. Yet on detail pages some functions—Back, Favorite, and Like—are on the top action bar. I propose an alternative access to these functions using the C-Swipe pattern.

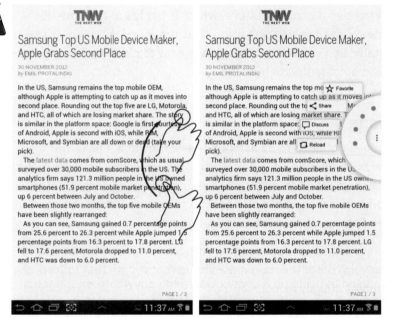

FIGURE 14.13: This wireframe shows a redesign of the Flipboard app with a C-Swipe menu.

A semicircular swipe with a thumb anywhere on the screen brings up a hidden menu containing the app's menu functions.

When and Where to Use It

The C-Swipe pattern is basically a complete replacement of the current Android action bar menu, so you can use it anywhere you might currently use an action bar. Particularly good candidates include tasks that today call for the lights-out mode of hidden Swiss-Army-Knife navigation, such as reading or browsing magazines.

Why Use It

The C-Swipe pattern has a number of important benefits:

- It is a highly immersive pattern that keeps the functionality hidden until it is needed, enabling 100 percent of the screen to be devoted to uniquely immersive touch experiences such as virtual reality.

- The C-Swipe pattern is unique because it can be used to create navigation anywhere on the screen, and the menu always shows up exactly where the hand is already positioned, so there is minimal strain and no need to change hand positions to access functionality.

- The C-Swipe pattern uses a unique gesture, which is not (as of the time this book is being written) used for anything else. Thus the menu is unlikely to be triggered accidentally, and when it is opened, it is easy to close by tapping anywhere on the screen outside the menu.

Other Applications

The C-Swipe pattern does not need to be activated near the edge. For larger touch screens that are tilted to enable the operator to remain comfortably standing, the C-Swipe navigation can be opened anywhere by making a swiping semicircular gesture with a thumb. Usually one of the other fingers, such as the index finger, must touch the screen first to support the compact gesture.

The C-Swipe pattern is unique because it comes with its own "natural" animated transition, which simply spins out the menu in a semicircular path that follows the movement of the thumb as closely as possible.

! Caution

The C-Swipe pattern is not easily discoverable. However, its discoverability can be boosted using an animated watermark or similar pattern. Stating something such as, "Swipe with your thumb anywhere on the screen," or playing the animated watermark several times in different places on the screen can help customers discover the gesture. After the initial discovery, the C-Swipe pattern quickly becomes familiar because it is so natural for a human hand.

Some people believe that the drawback of the C gesture is that it is ergonomically complex and does not translate into anything "real" such as scroll/pan. Other designers instead prefer larger gestures of the Windows Modern UI that use the entire arm to transverse the screen from left to right and top to bottom. Yet still other designers prefer another alternative for large displays: a special multitouch gesture, for example a five-finger tap or five-finger pinch. I disagree. Although each of these alternatives to the C-Swipe pattern holds a great deal of promise, I find the C-Swipe pattern to be the most natural, authentic, and economical of touch movements. Much testing with a broad range of people is needed to confirm this. However, one thing is clear: Regardless of the gesture, the concept behind a hidden menu is the correct one for the future of touch on larger tablets, so my biggest caution would be against ignoring this important trend.

Related Patterns

13.5 Pattern: Watermark

13.6 Pattern: Swiss-Army-Knife Navigation

INDEX